"In *Learning to Die in the Anthropocene*, Roy Scranton draws on his experiences in Iraq to confront the grim realities of climate change. The result is a fierce and provocative book."

—**Elizabeth Kolbert**, 2015 Pulitzer Prize winner and author of *The Sixth Extinction: An Unnatural History*

"Roy Scranton lucidly articulates the depth of the climate crisis with an honesty that is all too rare, then calls for a reimagined humanism that will help us meet our stormy future with as much decency as we can muster. While I don't share his conclusions about the potential for social movements to drive ambitious mitigation, this is a wise and important challenge from an elegant writer and original thinker. A critical intervention."

—**Naomi Klein**, author of *This Changes Everything: Capitalism vs. the Climate*

"We're fucked. We know it. Kind of. But Roy Scranton in this blistering new book goes down to the darkness, looks hard and doesn't blink. He even brings back a few, hard-earned slivers of light. . . . What is philosophy? It's time comprehended in thought. This is our time and Roy Scranton has had the courage to think it in prose that sometimes feels more like bullets than bullet points."

—**Simon Critchley**, editor of *The New York Times* Opinionator blog "The Stone," author of *Infinitely Demanding: Ethics of Commitment, Politics of Resistance*

"Roy Scranton gets it. He knows in his bones that this civilization is over. He knows it is high time to start again the human dance of making some other way to live. In his distinctive and original way he works though a common cultural inheritance, making it something fresh and new for these all too interesting times. This compressed, essential text offers both uncomfortable truths and unexpected joy."
—**McKenzie Wark**, author of *Molecular Red: Theory for the Anthropocene*

"An eloquent, ambitious, and provocative book."
—**Rob Nixon**, author of *Slow Violence and the Environmentalism of the Poor*

"Roy Scranton has written a howl for the Anthropocene—a book full of passion, fire, science and wisdom. It cuts deeper than anything that has yet been written on the subject."
—**Dale Jamieson**, author of *Reason in a Dark Time: Why the Struggle Against Climate Change Failed—and What It Means for Our Future*

Learning to Die
in the Anthropocene

Learning to Die
in the Anthropocene

REFLECTIONS
ON THE END OF A CIVILIZATION

Roy Scranton

City Lights Books

Grateful acknowledgment is made for permission to quote from the following:
The Epic of Gilgamesh: The Babylonian Epic Poem and Other Texts in Ak-kadian and Sumerian. Translation by Andrew George (London: Allen Lane, 1999). ©1999 Andrew George. Used by permission. All rights reserved.

"Fade," by Carrie Brownstein, Corin Tucker and Janet Weiss, ©2015 BMG Gold Songs/Songs Of Big Deal/Code Word nemesis (ASCAP). All rights administered by BMG Rights Management (US) LLC. Used by permission. All rights reserved.

Library of Congress Cataloging-in-Publication Data
Scranton, Roy, 1976-
 Learning to die in the Anthropocene : reflections on the end of a civilization / Roy Scranton.
    pages cm
 ISBN 978-0-87286-669-0 (paperback)
ISBN: 978-0-87286-670-6 (ebook)
1. Global warming. 2. Climatic changes. 3. Environmental degradation. 4. Nature—Effect of human beings on. 5. Climate change mitigation. I. Title.

 QC981.8.G56S33 2015
 303.49—dc23

                          2015022985

City Lights Books are published at the City Lights Bookstore
261 Columbus Avenue, San Francisco, CA 94133
www.citylights.com

CONTENTS

Dedicated to my brother, who taught me to remember the dead, and to Laura, who taught me to fight like hell for the living.

A free man thinks of death least of all things, and his wisdom is a meditation of life, not of death.
—Baruch Spinoza, *Ethics*, IV.67

COMING HOME

The knowledge of future things is, in a word,
identical with that of the present.

—Plotinus, *Enneads* IV.12

Driving into Iraq in 2003 felt like driving into the future.
We convoyed all day, all night, past Army checkpoints and
burned-out tanks, till in the blue dawn Baghdad rose from
the desert like a vision of hell: flames licked the bruised sky
from the tops of refinery towers, cyclopean monuments
bulged and leaned against the horizon, broken overpasses
swooped and fell over ruined suburbs, bombed factories,
and narrow ancient streets.

With "shock and awe," the US military had unleashed
the end of the world on a city of six million—a city about the
same size as Houston or Washington, D.C. Baghdad's infrastructure was totaled: water, power, traffic, markets, and
security fell to anarchy and local rule. The government had
collapsed, walls were going up, tribal lines were being drawn,
and brutal hierarchies were being savagely established. Over
the next year, the city's secular middle class would disappear,

squeezed out by gangsters, profiteers, fundamentalists, and soldiers.

I was a private in the United States Army. This damaged world was my new home. If I survived.

Two and a half years later, still in the Army but safe and lazy back in Fort Sill, Oklahoma, I thought I had made it out. Then I watched on television as Hurricane Katrina hit New Orleans. This time it was the weather that inspired shock and awe, but I saw the same chaos and collapse I'd seen in Baghdad, the same failure of planning and the same tide of anarchy. The 82nd Airborne Division took over strategic points and patrolled streets now under de facto martial law. My unit was put on alert and trained for riot control operations. The grim future I'd seen in Baghdad had come home: not terrorism, not WMDs, but the machinery of civilization breaking down, unable to recuperate from shocks to its system.

That future is not going away. According to Admiral Samuel J. Locklear III, head of the US Pacific Command, global climate change is the greatest threat the United States faces, more dangerous than terrorism, Chinese hackers, and North Korean nuclear missiles.[1] Upheaval from increased temperatures, rising seas, and climatic destabilization "is probably the most likely thing that is going to happen that will cripple the security environment, probably more likely than the other scenarios we all often talk about," he said. Thomas E. Donilon said much the same thing in 2014 as National Security Advisor, arguing that the "environmental impacts of climate change present a national security challenge."[2] James Clapper, Director of National Intelligence, told the Senate in 2013 that "Extreme weather events (floods, droughts, heat waves) will increasingly disrupt food and energy markets, exacerbating state weakness, forcing human migrations, and triggering riots, civil disobedience, and vandalism."[3] President Obama's 2010 *National Security Strategy*, the Pentagon's *2014 Quadrennial Defense Review*, and the Department of Homeland Security's *2014 Quadrennial Homeland Security Review* all identify climate change as a severe and imminent danger.[4] More recently, the Pentagon's *2014 Climate Change Adaptation Roadmap* warned: "Rising global temperatures, changing precipitation patterns, climbing sea levels, and more extreme weather events will intensify the challenges of global instability, hunger, poverty, and conflict. They will likely lead to food and water shortages, pandemic disease, disputes over refugees and resources, and destruction by natural disasters in regions across the globe."[5]

On the civilian side, the World Bank's 2013 report, *Turn Down the Heat: Climate Extremes, Regional Impacts, and the Case for Resilience*, and their 2014 follow-up *Confronting the New Climate Normal*, offer dire prognoses for the effects of global warming, which climatologists now predict will raise global temperatures 3.6 degrees Fahrenheit above preindustrial levels within a generation and 7.2 degrees Fahrenheit within 90 years.[6] As hotter temperatures liquefy glaciers and ice sheets from Greenland to Antarctica, all that melted ice flows into the sea: Some worst-case estimates suggest we might see seven or eight feet of sea level rise as soon as 2040.[7] The collapse of the West Antarctic ice sheet alone, already underway, will eventually raise sea levels by as much as twenty feet.[8]

As glaciers and ice sheets melt, so too will carbon and methane long frozen in seabeds and permafrost. As a greenhouse gas, methane is more than twenty times more powerful than carbon dioxide, and thousands of gigatons of the stuff lies locked under the oceans in clathrate hydrates, waiting to be released: "These solid, ice-like structures are stable only under specific conditions," writes oceanographer John Kessler, "and are estimated to contain a quantity of methane roughly equal in magnitude to the sum of all fossil fuel reservoirs on Earth."[9] Methane-rich sinkholes have appeared in Siberia and methane bubbles have been tracked leaking from the floor of the Arctic Ocean, possibly signaling the beginning of a massive planetary "belch" capable of generating catastrophic runaway greenhouse effects.[10] As geophysicist David Archer warns, "The potential for planetary devastation posed by the methane hydrate reservoir . . . seems comparable to the destructive potential from nuclear winter or from a comet or asteroid impact."[11]

We're fucked. The only questions are how soon and how badly. The Intergovernmental Panel on Climate Change's (IPCC) 2014 report on climate impacts cautions: "Without additional mitigation efforts beyond those in place today, and even with adaptation, warming by the end of the 21st century will lead to high to very high risk of severe, widespread, and irreversible impacts globally."[12] According to the World Bank, 2.7 degrees Fahrenheit of warming now appears inevitable, even if we were to stop emitting carbon dioxide (CO_2) worldwide right now.[13] Projections from researchers at the University of Hawai'i find us dealing with "historically unprecedented" climates as soon as 2047.[14] Climate scientist James Hansen, formerly with NASA, has argued that we face an "apocalyptic" future—a bleak view that is seconded by researchers worldwide.[15]

This chorus of Cassandras predicts a radically changing global climate causing widespread upheaval, and their visions of doom are backed by an overwhelming preponderance of hard data. Global warming is not the latest version of a hoary fable of annihilation. It is not hysteria. It is a fact. And we have likely already passed the point where we could have done anything about it. From the perspective of many policy experts, climate scientists, and national security officials, the concern is not whether global warming exists or how we might prevent it, but how we are going to adapt to life in the hot, volatile world we've created.

There is a name for this new world: the Anthropocene. The word comes from ancient Greek. All the epochs of the most recent geological era (the Cenozoic) end in the suffix "-cene," from *kainós*, meaning new. *Anthropos* means human. The idea behind the term "Anthropocene" is that we have entered a new epoch in Earth's geological history, one characterized by the advent of the human species as a geological force.[16] The biologist Eugene F. Stoermer and the Nobel-winning chemist Paul Crutzen advanced the term in 2000, and it has gained acceptance as evidence has grown that the changes wrought by global warming will affect not only the world's climate and biodiversity, but its very geological structure, and not just for centuries, but for millennia.[17] In the prophetic words of William Blake, written at the dawn of the carbon era more than two hundred years ago, "The generations of men run on in the tide of Time / But leave their destin'd lineaments permanent for ever and ever."[18]

The International Commission on Stratigraphy, the geologists responsible for driving the "golden spikes" that demarcate different geological periods, have adopted the Anthropocene as a term deserving further consideration, "significant on the scale of Earth history," and are discussing what level of geological time-scale it might be and at what date we might say it began.[19] Is it an "epoch" like the Holocene, or merely an "age" like the Calabrian? Did it start with the beginning of the Industrial Revolution, around 1800, or during the Great Acceleration in the middle of the 20th century? With the dawn of agriculture, 12,000 years ago, or on July 16, 1945, with the first atomic bomb?[20]

Whenever it began, it is the world we now live in. Within a few generations we will face average temperatures 7 degrees Fahrenheit warmer than they are today, rising seas at least three to ten feet higher, and worldwide shifts in crop belts, growing seasons, and population centers. Unless we stop emitting greenhouse gases wholesale now, humans will within a couple hundred years be living in a climate the Earth hasn't seen since the Pliocene, three million years ago, when oceans were 75 feet higher. Once the methane hydrates under the oceans and permafrost begin to melt, we may soon find ourselves living in a hothouse climate closer to that of the Paleocene-Eocene Thermal Maximum, approximately 56 million years ago, when the planet was ice-free and tropical at the poles. We face the imminent collapse of the agricultural, shipping, and energy networks upon which the global economy depends, a large-scale die-off in the biosphere that's already well under way, and our own possible extinction as a species. If *Homo sapiens* survives the next millennium, it will be survival in a world unrecognizably different from the one we have known for the last 200,000 years.

In order for us to adapt to this strange new world, we're going to need more than scientific reports and military policy. We're going to need new ideas. We're going to need new myths and new stories, a new conceptual understanding of reality, and a new relationship to the deep polyglot traditions of human culture that carbon-based capitalism has vitiated through commodification and assimilation. Over and against capitalism, we will need a new way of thinking our collective existence. We need a new vision of who "we" are. We need a new humanism—a newly philosophical humanism, undergirded by renewed attention to the humanities.

Admittedly, ocean acidification, social upheaval, and species extinction are problems that humanities scholars, with their taste for fine-grained philological analysis, esoteric debates, and archival marginalia, might seem remarkably ill-suited to address. After all, how will thinking about Kant or Frantz Fanon help us trap carbon dioxide? Can arguments between object-oriented ontology and historical materialism protect honeybees from colony collapse disorder? Are ancient Greek philosophers, medieval poets, and contemporary metaphysicians going to save Bangladesh from being inundated by the Indian Ocean?

Perhaps not. But the conceptual and existential problems that the Anthropocene poses are precisely those that have always been at the heart of humanistic inquiry: What does it mean to be human? What does it mean to live? What is truth? What is good? In the world of the Anthropocene, the question of individual mortality—What does my life mean in the face of death?—is universalized and framed in scales that boggle the imagination.[21] As environmental philosopher Dale Jamieson puts it, "The Anthropocene presents novel challenges for living a meaningful life."[22] Historian and theorist Dipesh Chakrabarty has claimed that global warming "calls us to visions of the human that neither rights talk nor the critique of the subject ever contemplated."[23] Whether we are talking about ethics or politics, ontology or epistemology, confronting the end of the world as we know it dramatically challenges our learned perspectives and ingrained priorities. What does consumer choice mean compared against 100,000 years of ecological catastrophe? What does one life mean in the face of mass death or the collapse of global civilization? How do we make meaningful decisions in the shadow of our inevitable end?

These questions have no logical or empirical answers. They cannot be graphed or quantified. They are philosophical problems *par excellence*. If, as Montaigne asserted, "To philosophize is to learn how to die," then we have entered humanity's most philosophical age, for this is precisely the problem of the Anthropocene.[24] The rub now is that we have to learn to die not as individuals, but as a civilization.

———————

Learning to die isn't easy. In Iraq, at the beginning, I was terrified by the idea. Baghdad seemed crazily dangerous, even though statistically I was relatively safe. We got shot at, mortared, and blown up by IEDs, but we wore high-tech ballistic armor, we had great medics, and we were part of the most powerful military the world had ever seen.[25] The odds were good that I would come home, maybe wounded, but probably alive. Yet every day I drove out past the wire on mission, I looked in my Humvee's mirror and saw a dark, empty hole.

"For the soldier death is the future, the future his profession assigns him," wrote Simone Weil in her 1939 meditation on war, *The Iliad, or the Poem of Force*. "Yet the idea of man's having death for a future is abhorrent to nature. Once the experience of war makes visible the possibility of death that lies locked up in each moment, our thoughts cannot travel from one day to the next without meeting death's face."[26] I recognized that face in the dark of my Humvee's mirror. Its gaze almost paralyzed me.

I found my way forward through an old book: Yamamoto Tsunetomo's 18th-century Samurai manual, the *Hagakure*, which advised: "Meditation on inevitable death should be performed daily."[27] I took that advice to heart, and instead of fearing my end, I practiced owning it. Every morning, after doing maintenance on my Humvee, I would imagine getting blown up, shot, lit on fire, run over by a tank, torn apart by dogs, captured and beheaded. Then, before we rolled out through the wire, I'd tell myself that I didn't need to worry anymore because I was already dead. The only thing that mattered was that I did my best to make sure everyone else came back alive.

To survive as a soldier, I had to learn to accept the inevitability of my own death. For humanity to survive in the Anthropocene, we need to learn to live with and through the end of our current civilization. Change, risk, conflict, strife, and death are the very processes of life, and we cannot avoid them. We must learn to accept and adapt.

The human psyche naturally rebels against the idea of its end. Likewise, civilizations have throughout history marched blindly toward disaster, because humans are wired to believe that tomorrow will be much like today. It is hard work for us to remember that this way of life, this present moment, this order of things is not stable and permanent. Across the world today, our actions testify to our belief that we can go on like we are forever: burning oil, poisoning the seas, killing off other species, pumping carbon into the air, ignoring the ominous silence of our coalmine canaries in favor of the unending robotic tweets of our new digital imaginarium. Yet the reality of global climate change is going to keep intruding on our collective fantasies of perpetual growth, constant innovation, and endless energy, just as the reality of individual mortality shocks our casual faith in permanence.

The greatest challenge the Anthropocene poses isn't how the Department of Defense should plan for resource wars, whether we should put up sea walls to protect Manhattan, or when we should abandon Miami. It won't be addressed by buying a Prius, turning off the air conditioning, or signing a treaty. The greatest challenge we face is a philosophical one: understanding that this civilization is already dead. The sooner we confront our situation and realize that there is nothing we can do to save ourselves, the sooner we can get down to the difficult task of adapting, with mortal humility, to our new reality.

Carbon-fueled capitalism is a zombie system, voracious but sterile. This aggressive human monoculture has proven astoundingly virulent but also toxic, cannibalistic, and self-destructive. It is unsustainable, both in itself and as a response to catastrophic climate change. Thankfully, carbon-fueled capitalism is not the only way humans can organize their lives together. Again and again throughout our history, we have shown ourselves to be capable of shedding maladaptive systems of meaning and economic distribution, developing resilient social technologies in response to precarity and threat, and transforming obsolete social practices into novel forms of life. Humanity's survival through the collapse of carbon-fueled capitalism and into the new world of the Anthropocene will hinge on our ability to let our old way of life die while protecting, sustaining, and reworking our collective stores of cultural technology. After all, our capacities to innovate and adapt depend on our being able to draw from our immense heritage of intellectual production, living and dead, exotic and close at hand: from the Iñupiat and from Islam, from Heraclitus and Zhuangzi, from the Torah and from the Buddha, from the *Federalist Papers* and from the *Communist Manifesto*. Carbon-fueled capitalism has given rise to a truly marvelous liberal multiculturalism, but if we are to survive its death throes, tolerance must mature into conservation and synthesis, grounded in a faith in human community existing beyond any parochial identity, local time, or single place.

The argument of this book is that we have failed to prevent unmanageable global warming and that global capitalist civilization as we know it is already over, but that humanity can survive and adapt to the new world of the Anthropocene if we accept human limits and transience as fundamental truths, and work to nurture the variety and richness of our collective cultural heritage. Learning to die as an individual means letting go of our predispositions and fear. Learning to die as a civilization means letting go of this particular way of life and its ideas of identity, freedom, success, and progress. These two ways of learning to die come together in the role of the humanist thinker: the one who is willing to stop and ask troublesome questions, the one who is willing to interrupt, the one who resonates on other channels and with slower, deeper rhythms.

The form this book takes is that of a story, but not a story about a person. Climate change is too big to be reduced to a single narrative, and the problems it presents us with demand that we transcend visually representative "picture-thinking" and work instead to create a sense of collective humanity that exists beyond any one place, life, or time. The story this book tells is of the human soul coming to know itself in its mortality. It begins in the deepest origins of our primal relationship with the Earth's climate, in Chapter 1: Human Ecologies, which traces that relationship up through our current moment and our contemporary predicament. In Chapter 2: A Wicked Problem, we consider that predicament. Carbon-fueled capitalism and its techno-utopian ideologues have promised infinite growth and infinite innovation, yet they have proven incapable of saving us from the disaster they have made. Various "solutions" to climate change have been offered, from carbon taxes to geoengineering, but none of them are likely to work. This chapter takes a look at the reasons why.

The global failure to address climate change is fundamentally a collective action problem, meaning it is a political problem. In Chapter 3: Carbon Politics, we consider how our collective failure to respond to climate change is an effect of the very structures of our political systems and the way that they are built around decentralized flows of oil and gas. The systems that structure our political desires and constrain our political will have a material history. As the human animal developed increasingly complex social technologies for producing power, from hunting bands tracking migrating herds of giant elk and mastodons to agricultural empires harvesting grain to fossil-fuel-burning global capitalism, we also developed increasingly complex technologies of collective life. As our technologies of producing power changed, so did our technologies for distributing and controlling it. Today, global power is in the hands of a tiny minority, and the system they preside over threatens to destroy us all. With this in mind, we turn back to the collective danger that carbon-fueled climate change poses, this time considered in terms of our primal human responses to existential threat: fight or flight. Facing the fear of death and the inevitability of conflict in the Anthropocene is the task of Chapter 4: The Compulsion of Strife. Progressivist belief in the infinite perfectibility of the human animal depends significantly on carbon-fueled capitalism's promises of infinite economic growth. Accepting our limits means coming to terms with our innate violence and our inescapable mortality.

By learning to die, though, we can connect with and open up new possibilities for the human future, as I argue in Chapter 5: A New Enlightenment. Through interrupting social circuits of fear and reaction, looking deep into the face of death, and cultivating our rich stocks of human cultural technology, from the *Epic of Gilgamesh* and the *Bhagavad-Gita* to imagined Anthropocene futures, we open up a human relationship to the universe in which we might live not as parasitic consumers, but as co-creators—a relationship in which we might learn to live as the very light from which all our power ultimately flows.

The crisis of global climate change, the crisis of capitalism, and the crisis of the humanities in the university today are all aspects of the same crisis, which is the suicidal burnout of our carbon-fueled global capitalist civilization.

The odds of that civilization surviving are negligible. The odds of our species surviving are slim. The trouble we find ourselves in will likely prove too intractable for us to manage well, if we can manage it at all. Yet as German philosopher Peter Sloterdijk observes: "It is characteristic of being human that human beings are presented with tasks that are too difficult for them, without having the option of avoiding them because of their difficulty."[28] We cannot escape our fate. Our future will depend on our ability to confront it not with panic, outrage, or denial, but with patience, reflection, and love.

Our choice is a clear one. We can continue acting as if tomorrow will be just like yesterday, growing less and less prepared for each new disaster as it comes, and more and more desperately invested in a life we can't sustain. Or we can learn to see each day as the death of what came before, freeing ourselves to deal with whatever problems the present offers without attachment or fear.

If we want to learn to live in the Anthropocene, we must first learn how to die.

ONE

HUMAN ECOLOGIES

Erþe toc of erþe erþe wyþ woh,
Erþe oþer erþe to þe erþe droh,
Erþe leyde erþe in erþene þroh,
Þo heude erþe of erþe erþe ynoh.

Earth took of Earth, Earth with woe,
Earth other Earth to the Earth drew,
Earth laid Earth in an Earthen trough,
Then had Earth of Earth Earth enough.

—Anonymous Middle English Lyric[29]

The first human beings appeared in tropical Africa around two hundred thousand years ago, evolving out of proto-human hominids, born into a world that was even then undergoing intense climatic transformation.[30] The Earth was on average about 3.5 degrees Fahrenheit colder than it is today, and would soon average between 5 degrees and 9 degrees Fahrenheit colder. The global climate was frigid, dry, and harsh. The northern hemisphere was mantled in vast ice sheets up to two and a half miles thick that reached as far south as Ohio, England, Germany, and northern China. Thirty percent of the Earth's surface was covered in glaciers. The oceans were more than 270 feet lower than they are now. This was the world in which we first evolved and learned to survive.

Our ancestors, omnivorous hunter-gatherers who traveled in small bands, lived in equatorial Africa for many thousands of years. But when temperatures warmed up about 135,000 years ago, reaching averages as much as 5 degrees Fahrenheit warmer than the present, our ancestors fled the tropics for more temperate climes, hunting their way into the then-verdant grasslands of the Sahara. The Earth had entered what geoscientists call an interglacial period, a regular period of warming that occurs between much longer periods of colder temperatures every hundred thousand years or so.[31] These periods are typically brief, geologically speaking, each lasting around ten thousand years.

So it was for our ancestors on the Sahara: After a few millennia, temperatures dropped, getting even colder than before (between 7 and 10 degrees Fahrenheit colder than the present), glaciers once again advanced toward the equator, and the oceans shrank. The Sahara lost the humid warmth that had kept it green, and our hunter-gatherer ancestors scattered back south to tropical Africa, east to the Nile valley and beyond, and north around the Mediterranean and into Europe, where they lived alongside, interbred with, and then wiped out our close cousin, the Neanderthal. Over a long winter lasting thousands and thousands of years, as minor climate variations shifted temperatures up and down slightly over decades and centuries, human hunting bands expanded across Asia, eventually crossing the Bering land bridge to the Americas.

Sometime in that period, between one hundred thousand and fifty thousand years ago, we developed the key social technologies that have proven our most substantial predatory advantages: culture and symbolic reasoning.[32] This advance was doubtlessly part of what helped us exterminate the Neanderthals and survive the harsh winters, and it also helped us turn the basic tool-use we had inherited from earlier hominids into the intricate technologies of the Cro-Magnon hunter: multilayer winter clothing, mammoth-bone houses, sewing needles, antler-tipped spears, religious rituals, and sophisticated tribal organization. Then, from about fifty thousand to about ten thousand years ago, Upper Paleolithic peoples experienced a veritable cultural explosion. The artifacts they left show technologies more advanced in almost every way than those of their predecessors, more varied, more elaborate, more refined, and more beautiful, with some tools even seemingly designed to be purely aesthetic in function—which is to say, works of art.[33]

About fifteen thousand years before Hurricane Katrina, the planet started to warm up again, entering another interglacial period, with the most intense and rapid warming happening around 11,000 BCE. Humans began developing villages, basic animal husbandry, and more deliberate systems of gathering. A brief, localized cold snap called the Younger Dryas, caused by glacial meltwater spilling into the Atlantic Ocean and shutting off the Gulf Stream, brought frigid temperatures to Europe and drought to southwest Asia for a thousand years.[34] According to archaeologist Brian Fagan, it was almost certainly this drought that drove the people in a small village on the Euphrates, who had until then depended for their sustenance on hunting desert gazelles, harvesting nuts, and gathering wild grasses, to begin deliberately cultivating rye, lentils, and a grain called einkorn.[35] By 9000 BCE, after the Younger Dryas had ended and the Gulf Stream had switched back on, the agricultural revolution had begun. Neolithic humans hunted, herded, gathered, and farmed from Europe to South America, thriving in the warm and mild climate of what is now called the Holocene.

A few thousand years later, the Laurentide ice sheet in northern Canada collapsed, causing a rapid rise in sea levels, and, as had happened with the Younger Dryas, shutting off the Atlantic Gulf Stream. Cold, dry conditions descended on Europe and southwest Asia. This cold drought lasted four centuries before the Gulf Stream switched on again. In the marshy confluence of the Tigris and Euphrates rivers, farmers began constructing irrigation canals to control flooding and increase crop yields. Villages grew into towns, and as the people living in these new towns struggled to deal with the complex, difficult group effort required to construct and maintain levee systems, keep records of floods and farm yields, and bring in the harvest, they began to develop refined divisions of labor, hierarchical political structures, sophisticated religions, and writing. Temples and marketplaces were built, traders carried goods from one town to another, and priests accumulated power as they hoarded knowledge of weather patterns and seed growth. The settlements of Ur, Eridu, and Uruk grew from clusters of villages to immense cities of tens of thousands. Through a complex interplay of droughts, population growth, and the increasing centralization of agriculture, human society based in and around cities—what we call "civilization"—emerged in the fertile crescent of land arcing from present-day Lebanon through Syria and Iraq to the Persian Gulf. From approximately 3100 to approximately 2200 BCE, Sumerian and then Akkadian kings ruled a vibrant collective form of life that stretched across Mesopotamia. But when a three-hundred-year-long drought hit the region, that vibrant empire fell apart.[36] Mighty Uruk stood desolated.

From ancient Uruk to modern-day Iraq we span about five thousand years, including nearly all of recorded human history—the Greeks, the Romans, the Tang dynasty, the Mongolian khanate, World War II, the invention of the cellphone, and all seven seasons of *Mad Men*. If human existence on Earth were a day, our approximately five millennia of recorded history would take up the last half hour before midnight. Throughout 99.9 percent of humanity's two hundred thousand years on Earth, the average planetary temperature never rose above 61 degrees Fahrenheit and carbon dioxide concentrations never went above 300 parts per million (ppm). Nearly all of our energy came from photosynthetic processes: most of our fuel was plant food, for ourselves directly, through the animals we used and preyed on, or through biomass like wood, with some use of water and air power through technologies like mills and sails, and negligible use of coal and oil.

Then, in 1781, James Watt invented the continuous-rotation steam engine. Suddenly power was portable, independent of living beings or natural forces, and able to run continuously. The steam turbine offered a vast improvement in energy production over wind, water, and animal labor, but it needed dense, hot-burning fuel for maximum output. Luckily for Watt, there happened to be loads of the stuff all over England: fossilized carbon in the form of coal.

Industrial coal changed everything. For the last two hundred years, just about one tenth of one percent of human existence, most of our energy has come not from direct photosynthesis but from stored carbon energy in fossil fuels. Switching from a photosynthetic-based energy economy to a carbon-based energy economy increased human wealth beyond what anyone could have possibly imagined, raising the overall standard of living across the world through such technologies as diesel-fueled tractors, Haber-process nitrogen-fixed fertilizer, Bessemer steel, railroads, steamships, airplanes, electric power plants, plastics, the internal combustion engine, and the automobile. It also began a massive transformation of the physical systems regulating life on Earth. By transferring millions of tons of carbon from the ground into the air, we have wrought profound changes in the Earth's climate, biosphere, and geology. Average atmospheric CO_2 levels have rocketed from 290 ppm to over 400 ppm, a level the planet hasn't seen in more than two million years. At the same time, methane (CH_4) levels have increased from 770 parts per billion to more than 1,800 parts per billion, the highest concentration of atmospheric methane in at least eight hundred thousand years.[37] These changes have disrupted the climate patterns regulated by the Earth's orbit around the Sun and will continue to disrupt them for thousands of years.

Our planet can sustain life because when energy from the Sun strikes the Earth, some of it is trapped in the atmosphere as heat. There are several greenhouse gases that help this happen, including carbon dioxide, methane, water vapor, and nitrous oxide. These gases are integral to the planet's complex geophysical homeostasis. That homeostasis has shifted, over millions of years, back and forth between "greenhouse" and "icehouse" states. During the last major greenhouse state, in the Eocene, the planet was ice-free, tropical from pole to pole, up to 20 degrees Fahrenheit warmer than it is now, and had CO_2 concentrations up to ten times those of today. The oceans were 300 feet higher. Large reptiles and dwarf mammals ranged through lush forests: a forty-foot-long snake that weighed more than a ton, a tiny horse the size of a dog, sleek feline predators, and lemur-like primates. Crocodiles and palm trees thrived along the Arctic Circle. That was fifty million years ago. Since then, the Earth has cooled, and it has been in an "icehouse" state for more than two and a half million years. We have very likely brought that state to a premature end.

For the first sixty thousand years of *Homo sapiens*' life on Earth, global temperatures 5 to 9 degrees Fahrenheit colder meant an ice sheet covering what is now Chicago and New York. That same amount—5 to 9 degrees Fahrenheit—is about as much as the planet is expected to warm up over the next few generations, and it doesn't sound like a lot. After all, the temperature changes more than that every day, and frequently much more, depending on season and locale. But when you're talking about planetary averages, those differences are enormous. A future 5 to 9 degrees Fahrenheit warmer will mean the Arctic Ocean will be ice free in the summer. Mountain glaciers will all but disappear, and with them, skiing, snowpack and a great many freshwater streams. Freak weather will play havoc with agricultural systems, plant and animal habitats, and human infrastructure. We'll have to contend with more extreme temperature fluctuations, more humidity, more and more intense rainfall, stronger and longer-lasting storms, severe droughts, and unpredictable changes in formerly reliable climate dynamics, such as the jet streams, the El Niño southern oscillation, and the Gulf Stream. Coral reefs will go extinct, along with countless other species caught in ecosystems changing too swiftly for them to adapt to or migrate out of.

More important, warming of 5 to 9 degrees Fahrenheit will eventually lead to sea levels 90 to 200 feet higher. No one is sure how quickly that will happen: If the ice melts slowly, we might only see a few feet of sea level rise by 2100. If the ice melts quickly and the Greenland ice sheet collapses, we could witness seas 20 to 30 feet higher within decades. Ice sheets are already melting faster than models have predicted, there is evidence that they have broken apart very quickly in the past, and ice melt is a feedback phenomenon, meaning that the more ice that melts, the faster the remaining ice melts. These factors mean that a rapid, unpredictable rise in sea level is all too possible. According to James Hansen, "the empirical data show us that natural ice sheet disintegration can be rapid, at rates up to several meters of sea level rise per century."[38] Whether it happens slowly or quickly, sea level rise will be disastrous for modern civilization: hundreds of millions of people living in low-lying coastal cities will be threatened not only by floods, but also by the increased storm surges resulting from the combination of higher seas, a wetter climate, and stronger storms.

It gets worse. Global warming of 5 to 9 degrees Fahrenheit depends on "business as usual," but there are very good odds that business will not go as usual. Growing populations and surging carbon consumption, particularly in China and India, will mean that the amount of carbon waste being produced is likely to *increase* over the next eighty-five years, as it has increased over recent decades. Feedback dynamics in the global climate system will likely raise temperatures even faster. Melting permafrost in Canada and Siberia will significantly increase atmospheric carbon dioxide and potentially increase warming by up to 80 percent.[39] And, as mentioned before, methane hydrates frozen in permafrost and locked in sediments at the bottom of the ocean could "belch," superheating the Earth and likely making it uninhabitable for the primate *Homo sapiens*.

Our hominid ancestors evolved during a period of general planetary cooling, and humans themselves evolved in a glacial climate colder than the one we live in today. Civilization as we understand it developed during what has been an unusually long and mild interglacial period, beginning around 10,000 BCE and continuing into the recent past. After the icy millennia of the late Pliocene, the Holocene was a kind of Eden, and being the clever, adaptable animals that we are, we took advantage of it. Human civilization has thrived in what has been the most stable climate interval in 650,000 years.[40] Thanks to carbon-fueled industrial civilization, that interval is over.

TWO

A WICKED PROBLEM

> I have seen this swan and
> I have seen you; I have seen ambition without
> understanding in a variety of forms.

> —Marianne Moore, "Critics and Connoisseurs"

In Iraq today, as in Uruk four thousand years ago, human beings live at the mercy of a changing climate. Decreased rainfall and diminished snowpack in the mountains mean that the Tigris and Euphrates are drying up, a problem exacerbated by new dams, increased water use, and water diversion in Turkey, Iran, and Syria. Drought, Saddam Hussein's draining of the Mesopotamian Marshes, and the devastation and abandonment of farmland due to years of war, neglect, and neoliberal economic policies favoring foreign imports over local produce have increased desertification, which in turn is unleashing punishing dust storms on Iraq's cities and crops.[41]

There are, of course, many differences between Uruk's distant collapse and Iraq's ongoing crisis. One of the most important is that Iraq's situation is man-made, while Uruk's was not. Another is that in 2200 BCE, the only things that beleaguered Sumerians could do in response to a changing climate were ration their grain and pray. Today, national, regional, and local governments worldwide, in cooperation with international bodies such as the IPCC, the United Nations Framework Convention on Climate Change (UNFCCC), the World Bank, the International Energy Agency (IEA), and the World Trade Organization (WTO), confront the problem of global warming with tremendous resources, the knowledge of thousands of highly trained scientists and engineers, and the support of hundreds of thousands of dedicated activists and concerned citizens. Yet for all that, we seem no more capable than were the people of Uruk when it comes to rescuing ourselves from imminent catastrophe.

The scientific study of climate change goes back to the early nineteenth century, when geologists and naturalists struggled to make sense of evidence suggesting that much of the Earth had once been covered in glaciers, and the science developed as physicists and chemists sought to understand the composition and mechanics of the Earth's atmosphere. The Swedish scientist Svante Arrhenius demonstrated the close relation between carbon dioxide levels and atmopheric temperature in 1895, theorizing what we now understand as the greenhouse effect and suggesting that widespread coal burning might increase global temperatures.[42] By the 1950s and 1960s, the effects of industrial pollution on the global climate were being studied by many scientists, among them Charles David Keeling, whose graph measuring carbon dioxide at the Mauna Loa Observatory in Hawai'i, the now-famous "Keeling Curve," showed clearly that atmospheric CO_2 was increasing. Over the next thirty years, evidence for man-made global warming grew, and by the late 1980s a scientific consensus had been established.

In 1988, James Hansen, then director of the National Air and Space Administration's Insitute for Space Studies, testified before the US Senate that the Earth was definitely warming, and "that it was 99 percent certain that the warming trend was not a natural variation but was caused by a buildup of carbon dioxide and other artificial gases in the atmosphere."[43] The Intergovernmental Panel on Climate Change was founded that same year to report and advise the United Nations on the problem of climate change, and the United Nations Framework Convention on Climate Change was established in 1992, committing its signatories to stabilizing global greenhouse gas emissions at a safe level. Every member nation of the UN signed the UNFCCC treaty in 1992 and most had ratified it by 1995, but the commitments they made came with no clear objectives, no viable mechanism for monitoring whether objectives were achieved, and no binding authority to enforce compliance.

In the decades since, while the almost two hundred nations committed to the UNFCCC have worked out individual emissions targets, they have not come to any agreement on monitoring or enforcement. Conference after conference has sunk under its own weight as a lack of accountability, intransigence from the US, China, and India, outsized goals set with no realistic plan for achieving them, bickering, and global power politics have led to failure after failure. Meanwhile, global greenhouse gas emissions have increased 35 percent since 1990, driven primarily by waste carbon dioxide from expanding energy consumption in North America and Asia.[44]

The only sure way to keep global warming from accelerating out of control would be to stop dumping waste carbon dioxide immediately. In the stern words of the IPCC: "Climate change can only be mitigated and global temperature be stabilized when the total amount of CO_2 is limited and emissions eventually approach zero."[45] With just the CO_2 in the atmosphere and oceans today, we are already set for at least 2 to 3 degrees Fahrenheit warming above pre-industrial levels, and it might be more like 5 or 6 degrees Fahrenheit. Any more CO_2 we put in from now on (by starting a car, for instance, or charging a phone) is only going to amplify that. So even if we banned dumping CO_2 right now, this very instant, we would still be facing serious climate impacts for centuries. Unfortunately for us, given the realities of global politics, a comprehensive, enforceable, worldwide ban on CO_2 is sheerest fantasy.

But what about other solutions? What about mitigation? What about decarbonizing our economy, replacing coal and oil with renewable energy or nuclear power? What about a carbon tax? What about cap-and-trade, carbon capture and sequestration, carbon extraction, and geoengineering? Might these strategies help us end, reduce, or at least mitigate our CO_2 emissions before we hit a tipping point and it's too late?

Ending our reliance on carbon-based fossil fuels—

decarbonizing the global economy—would be the most reliable path to limit and eventually stop dumping waste CO_2. The problem is that global decarbonization is effectively irreconcilable with global capitalism. Capitalism needs to produce profit in order to spur investment. Profit requires growth. Global economic growth, even basic economic stability, depends on cheap, efficient energy.

Decarbonizing the global economy without a replacement energy source would mean turning off approximately 80 percent of our power, causing a worldwide economic meltdown that would make the Great Depression look like a sluggish sales season. While not nearly as dire, worldwide decarbonization *with* replacement energy still looks pretty unpalatable. The most reliable studies suggest that even stabilizing CO_2 at a relatively low but still unsafe level would require long-term economic austerity. According to the Potsdam Institute for Climate Impact Research, stabilizing carbon dioxide levels at 450–500 ppm (which is 100–200 ppm *over* the upper limit for keeping warming anywhere near 3.6 degrees Fahrenheit [2 degrees Celsius]) calls for slowing and probably even contracting the global economy indefinitely, basically extending the Great Recession into the indefinite future.[46] No population on the planet today is going to willingly trade economic growth for lower carbon emissions, especially since economic power remains the key index of global status.[47] The political paroxysms of forced austerity rocking countries across Europe today are only a taste of what we would have to look forward to under a carbon-austerity regime.

Offering a glimmer of hope, the IPCC's 2014 report on mitigation argues that we can avoid the worst of global warming with what would be only a slight decrease in global economic growth—about .06 percent.[48] The report claims that shifting investment from oil and coal production to research and development of renewable energies, nuclear power, and carbon capture and sequestration could make it possible to decarbonize within thirty or forty years with only a slight cost to global gross domestic product. The report may be right, but the IPCC's numbers are speculative, not predictive: The renewable energy and carbon capture technologies the report's numbers depend on are still emerging, and we don't know yet whether they can work on a large enough scale to make a difference, or how much more they might cost than carbon fuels do now. As well, critics have pointed out a variety of problems with the IPCC's numbers on decarbonization, ranging from excessive political influence affecting data to biases in models underestimating economic impacts.[49]

Adding to our troubles, it's no simple thing to completely renovate worldwide energy infrastructure that has taken many years to build. Vaclav Smil, one of the world's leading energy analysts, observes: "There are five major reasons that the transition from fossil to nonfossil supply will be much more difficult than is commonly realized: scale of the shift; lower energy density of replacement fuels; substantially lower power density of the renewable energy extractions; intermittence of renewable flows; and uneven distribution of renewable energy resources."[50] It would take decades to develop and implement new systems of carbon-free or carbon-minimal energy infrastructure, if it's even possible, and we don't have decades.[51] Even 2035, a mere twenty years from now, will be too late. It's very likely already too late now.

Another major problem we face in decarbonizing and rebuilding global energy infrastructure is that the prime candidates for clean renewable power, solar and wind, are in themselves not reliable enough to supply the baseline energy we need to keep our lights burning, our EKGs and medical respirators beeping and pumping, and our servers crunching data. "The central and deeply intractable fact about electricity," writes journalist and historian Gwynne Dyer, "is that it must be generated at the very moment it is used. This poses a major problem for those in charge of supplying a society with electricity, since . . . the demand can vary as much as threefold from a mild weekend day in summertime to a cold midwinter evening."[52] The debilitating issue with solar and wind power is that they add to the problem of variable demand the problem of variable supply: they depend on the weather, which changes dramatically from day to day and hour to hour. Dyer writes:

> The managers of the big grids, who already have to cope with wildly fluctuating levels of demand, are now being asked to deal with uncontrollably fluctuating levels of supply as well, and they can only go so far. It hasn't come up much in public yet, because only Germany and Denmark have

approached even 20 percent renewables in their electricity-generation mix [recently as much as 27 percent], but most grid managers are very unhappy about going beyond that level of renewables in the system.[53] With better programmes for predicting short-term wind fluctuations they might be persuaded eventually to go up to 50 percent, but beyond that . . . you simply cannot go unless you are prepared to accept periodic collapses of the entire grid when the wind drops.[54]

Many critics question whether renewable energy will *ever* be a feasible alternative to carbon fuels, pointing out that its cost-to-energy ratios are simply too high. Even James Hansen, one of the most outspoken scientists on the urgent need to address global warming, remains skeptical: "Most energy experts agree that, for the foreseeable future, renewable energies will not be a sufficient source of electric power. There is also widespread agreement that there are now just two options for nearly carbon-free large-scale base-load electric power: coal with carbon capture and storage, and nuclear power."[55]

As Hansen wrote with three other scientists in a controversial open letter in 2013, "Renewables like wind and solar and biomass will certainly play roles in a future energy economy, but those energy sources cannot scale up fast enough to deliver cheap and reliable power at the scale the global economy requires. While it may be theoretically possible to stabilize the climate without nuclear power, in the real world there is no credible path to climate stabilization that does not include a substantial role for nuclear power."[56] Nuclear fission, however, presents significant and perhaps insurmountable waste issues, and in order to reduce carbon emissions enough to keep CO_2 below 450 ppm, the world would have to build more than 12,000 nuclear power plants in the next 35 years, "about the same as one new plant coming online every day between now and 2050."[57] That seems unlikely, especially since disasters such as Three Mile Island, Chernobyl, and Fukushima remain forbidding reminders of the risks nuclear power poses.

And then there's the question of how the nations of the world would even go about implementing global decarbonization if they all agreed to do it. National carbon taxes strict enough to actually decrease carbon use have been implemented in a few countries, but even so, global emissions have continued to rise, and we have yet to solve the problem of enforcement: Who is going to make the United States, China, Russia, and India pay more for their coal and oil? There would necessarily have to be some kind of international agreement on the price of carbon, which seems unlikely, given that scientists and economists in the United States and the United Kingdom—both wealthy nations with shared economic interests and a long history of CO_2 emissions—disagree substantially, estimating the costs of carbon variously "from a few to several hundred dollars per tonne."[58] As economist Michael Grubb describes the problem:

The holy grail of an economic cost-benefit calculation applied to climate change is to establish the "social cost of CO_2 emissions"—the damage inflicted by emitting each tonne. As shown, the calculation turns out to be just as uncertain as everything else—in fact more so. It does not provide an objective answer in a world of conflicting views, but unavoidably reflects back the assumptions and values that people bring to the table. Climate policy will have to contend with "incommensurable benefits estimates." Even the world's top economists have accepted that they cannot reach an agreement about how to tackle the problem, even on the single component of how to weight impacts over time, let alone other dimensions.[59]

Populations and countries in economic competition, with wildly unequal investments in fossil fuel reserves and facing wildly unequal impacts from the changing global climate, are highly unlikely to come to an agreement on how much a ton of CO_2 should cost, thus making it improbable that strict, universal regulatory taxes on carbon will ever be established. If this seems unduly pessimisstic, consider that neither the US, Russia, nor China—together responsible for around 48 percent of the world's carbon dioxide emissions—have a national carbon tax, and that two of those nations are fossil fuel exporters who depend on the growing energy markets of the third. Further, carbon taxes have faced intense opposition not only from fossil fuel producers, but also from voters: most notably, Australia became famous in 2014 as the first country in the world to repeal a national carbon tax.

Instead of such taxes, many people advocate decarbonization through a global "cap-and-trade" mechanism similar to the one that helped solve the problem of acid rain.[60] In a cap-and-trade scheme, emissions limits are established and gradually tightened, while emitters are allowed to trade emissions credits among themselves to minimize the costs of compliance. But Thomas Schelling, a Nobel-winning economist and strategic systems theorist, makes a strong case that carbon trading simply will not work. While he "believes in the essentiality of incentives, in clearly defined obligations, and in the virtues of trading," Schelling "cannot imagine such a regime for carbon emissions."[61] His main reasons are straightforward. First, "any serious regime would have to allocate emission rights over many decades, not just a decade at a time but cumulatively," and there is no way of reaching agreement on what emissions limits should be over the next hundred years or on deciding the costs of exceeding them. Second, "it would be almost impossible to determine, during the first half-century or so, whether a nation was on target to meet its ultimate cumulative limit." Third, enforcement would require an independent body able to punish poor and rich countries alike, and it is hard to see how such an enforcement body would function or from where it would derive its authority. As Schelling notes, "there is no historical example of any international regime that could impose penalties on a scale commensurate with the magnitude of global warming. . . . Nothing like this has ever existed and it is even hard to conceive."[62] Consider the fact that human trafficking, genocide, torture, the use of chemical weapons, and wars of aggression have all been banned by international agreements, yet such crimes keep happening and are often committed by the leading signatories of the very treaties that ban them. It seems irresponsible to expect things to be any different when it comes to dumping carbon.

Since we're not likely to regulate our carbon dioxide emissions away, maybe we should bury them. Carbon capture and sequestration (CCS) has been proposed as a technological solution that would do just that. With CCS, you put pipes on top of smokestacks to collect waste CO_2, then bury it all somewhere. The trouble is that CCS is a new and expensive technology that stands little chance of being developed on a global scale quickly enough to make a real difference. According to the International Energy Agency, robust enough CCS development to keep global warming below 7.2 degrees Fahrenheit would take global investment of *at least* 5 to 6 billion dollars a year. As of 2013, cumulative global investment in CCS technologies has only been averaging about 2 billion dollars a year.[63] The IEA's 2009 CCS roadmap called for 100 CCS projects to be developed by 2020, but as of 2013, only four had been completed, with nine more under construction. The four functioning projects had cumulatively stored about 50 million tons of carbon, while we dumped 200 times that amount of CO_2 into the atmosphere in 2012 alone.[64] As more recent IEA reports have shown, progress on carbon capture and sequestration does not inspire confidence. The IEA explains: "Because markets do not value the public benefits of CCS demonstration and the benefits cannot be captured in full by early adopters, there is currently little commercial incentive for private entities to invest in CCS."[65] Absent government incentives and regulation, there's no reason for anyone to expect that carbon capture and sequestration will be developed or implemented on a meaningful scale, and as long as coal and oil companies remain politically powerful, there's no reason to expect robust enough government action in time for CCS to be a viable response to global warming.

Pulling carbon dioxide out of the air has also been proposed as a technological fix to address emissions, though it too would be tremendously expensive and, like capture and sequestration, has so far remained underdeveloped. As Roger Pielke Jr. points out, "Presently, there are no experimental data on the complete process of air capture, especially at scale, to demonstrate the concept, its energy use, and the engineering costs."[66] We should be researching and developing such technologies, of course, while recognizing that they are far from being quick, easy solutions and may not ever be viable. The time for a speculative, miraculous, last-minute, long-shot techno-fix is well over. We are living *right now* in the midst of a global climate emergency and social crisis that demands immediate response and long-term adaptation. We must prepare for the coming storm—not in thirty or forty years, but today.

Some people want to believe we can hold off the storm by artificially cooling the planet with a crude process often disingenuously dignified by the term "geoengineering," otherwise known as "Solar Radiation Management."[67] The most technologically feasible plan is simple: we lay a blanket of sulfur in the stratosphere to increase the Earth's reflectivity, or albedo, thus shining back more of the Sun's light into space and thereby cooling the planet. While cheap enough that a small country or even large corporation could make it happen, this plan has serious flaws. Stratospheric sulfur aerosol cooling is dangerous, because of the toxicity of the sulfur, because it could degrade the ozone layer, and because the resulting increased albedo would make us more vulnerable to fallout from volcanic eruptions. If the atmosphere were already thick with particulate sulfur, additional aerosols from an eruption like Pinatubo or Eyjafjallajökull could force a sudden massive cooling like the one that happened when Mount Tambora exploded in 1816, the so-called "Year Without a Summer." Such an event would be catastrophic for agricultural yields and likely cause widespread famine. We also don't know how such deliberate dumping of sulfur into the atmosphere would further impact climatic changes already happening because of warming: clouds darkened with sulfur may absorb more heat, and particulate sulfur drifing down onto ice and snowfields may worsen the "dark snow" effect that is helping melt glaciers in Greenland and elsewhere. Finally, and most seriously, sulfur aerosol cooling simply doesn't work over the long term. Reflecting more light back into space deals with the symptom (warming) without addressing the cause (greenhouse gases). Sulfur aerosols are heavy and would tend to settle back on the

Earth's surface (and on us and our crops), so they would have to be continually dumped into the upper atmosphere, a situation that sets us up for sudden runaway heating if we should ever stop pumping up the sulfur—if there were another world war, for instance.

Global warming is what is called a "wicked problem":[68] it doesn't offer any clear solutions, only better and worse responses. One of the most difficult aspects to deal with is that it is a collective-action problem of the highest order. One city, one country, even one continent cannot solve it alone. Any politician who honestly and frankly worked to detach her nation's economy from oil and coal would not survive in any kind of democratic or oligarchic government, because the rigorous austerity necessary to such an effort would mean either economic depression and poverty for most of her constituency, a massive redistribution of wealth, or both. Moreover, any leader who forced her country to accept the austerity and redistribution necessary to end its dependence on cheap carbon would also be forcing her country into a weak and isolated position politically, economically, and militarily. The entire world has to work together to solve global warming, yet carbon powers the world's political machinery and shapes our current form of collective life. It's coal and oil that we have to thank for connecting the many nations of the world into one tight, integrated economy. Without the information, energy, and transportation infrastructures built and sustained with carbon, there wouldn't be any global civilization to try to save.

THREE

CARBON POLITICS

From low to high doth dissolution climb,
And sink from high to low, along a scale
Of awful notes, whose concord shall not fail.

—William Wordsworth, "Mutability"

When a honeybee colony needs to find a new home, it sends out wave after wave of scouts to search for a new hive site.[69] When the scouts return, they dance for the other bees. Each scout's dance communicates a possible location for the colony's future. As new waves of scouts go out and return, they align themselves with one dance or another, depending on what they've found. Soon, in a vast rollicking caucus, masses of bees are all dancing to a variety of distinct rhythms, each dance offering a different vision of tomorrow. One dance may be for a nearby oak, another for a distant elm; one dance offers an ambitious journey, another is more conservative. Over time, a single dance grows more and more popular, until a majority of the bees are doing it. The swarm has made its decision and takes flight.

Politics, whether for bees or for humans, is the energetic distribution of bodies in systems. This is where the ideas of the vote, the town hall meeting, and the public debate get their power: humans come together to resonate on one frequency or another. Arrangements of bodies in systems don't arise from ideal notions of how governance should work, but rather emerge out of the vibrating bodies themselves, the systems they inhabit, and the interactions between the two.

The key is energy: energy production and social energetics. Just as a beehive is structured around the production of honey, so are human societies structured around labor, horses, wheat, coal, and oil. How bodies harvest, produce, organize, and distribute energy determines how power flows, shaping the political arrangements of a given collective organism behind whatever ideologies the ruling classes may use to manufacture consent, obscure the mechanisms of control, or convince themselves of their infallible omniscience.

Humanity has undergone three major revolutions in the political structures of energy production in the past 200,000 years: the Agricultural Revolution, the Industrial Revolution, and the Great Acceleration (the transition from coal to mixed fossil fuels). The Agricultural Revolution shifted human social organization from the pack to the herd, from nomadic life to sedentary life, inaugurating politics as we understand it (meaning the life of the *polis*, the city, urban existence, "civilization"). With the advent of the farm and the city, humans no longer followed migrating energy stocks, tracking the sweeping herds of ungulates that were once a primary source of food, but cultivated stocks in place:

sheep, wheat, einkorn, dates. The distribution of bodies in fields, pastures, canals, and cities gave rise to new forms of social energetics that superseded tribes and tribal confederations. In ancient Uruk, despotism and empire emerged as political technologies to deal with the intricate demands of flood-based irrigation: single farmers or clans couldn't manage the intensive mass labor required to dredge canals and harvest grain at critical moments in the growing season, so centralized, absolutist systems and ideologies were invented to control production. Elsewhere, as agricultural technologies spread to fertile areas less dependent on annual floods and complex irrigation, less centralized political technologies were developed (such as feudalism).

Around 12,000 years after the invention of agriculture, the Industrial Revolution shifted human social organization from photosynthetic stocks to fossilized carbon stocks. This freed us from our dependence on plant and animal energy and opened up incredible new flows of power, while the organization of bodies around coal pits, railroads, and teeming conurbations in the 19th and early 20th centuries gave rise to mass social democracy, state nationalism, and industrial capitalism. As Timothy Mitchell argues in his book *Carbon Democracy*, the ability of tenacious, highly organized coal miners to interrupt energy flows gave them significant leverage in what had previously been essentially feudal and absolutist systems.[70] Rulers were forced to listen to workers, because coal miners at the forefront of the labor movement could interrupt the operations of an entire country. Through labor unions, general strikes, and sabotage, sustained by the ability of coal miners, railroad workers, teamsters, and long-shoremen to paralyze national economies, "working people in the industrialized West acquired a power that would have seemed impossible before the late 19th century." Mitchell writes:

> Workers were gradually connected together not so much by the weak ties of a class culture, collective ideology, or political organization, but by the increasing and highly concentrated quantities of carbon energy they mined, loaded, carried, stoked, and put to work. The coordinated acts of interrupting, slowing down, or diverting its movement created a decisive political machinery, a new form of collective capability built out of coalmines, railways, power stations, and their operators.[71]

That collective capability waned with the Great Acceleration in the middle of the 20th century, as industrial societies shifted from reliance on coal to the mixed use of coal, oil, and natural gas. Unlike a coal-based economy, which relies on raw labor in numbers, oil and gas production requires relatively few workers. Labor retained vestigial power for decades, but the realignment of energy flows from solid coal to liquid petroleum and natural gas severely weakened the effective political power coal miners and their allies could leverage, substantially undermining mass social democracy as a technology of power.

Liquid carbon stocks come to us through decentralized

networks managed by small crews of highly trained technicians and owned by a handful of corporations, nations, and individuals. Coal must be physically dug out of the ground and transported by fixed rail lines to distribution centers, but oil and gas are mechanically pumped by pipeline from remote wells to ports, where the liquids are loaded onto tankers that can be sent practically anywhere and rerouted in transit with ease. As Mitchell explains, "whereas the movement of coal tended to follow dendritic networks, with branches at each end but a single main channel, creating potential choke points at several junctures, oil flowed along networks that often had the properties of a grid, like an electricity network, where there is more than one possible path and the flow of energy can switch to avoid blockages or overcome breakdowns."[72] Populist movements that used to be able to organize around the centralized flows of coal civilization are all but powerless when it comes to interrupting the much more flexible flows of oil and gas (isolated high-profile cases like the Keystone XL pipeline notwithstanding).

Growing from and resonating with the flows of material power that sustain them, our political arrangements today are collective organisms of consuming bodies in decentralized systems managed by technicians for the profit of the few. At the top, there is an oligarchy of owners who control the lion's share of world energy production. They rule through a technocratic administrative class of operators and politicians, and run mass media elections to jockey for control amongst themselves and manufacture ritual consent. Most people participate if at all as consumers, watching the election games and voting for one of the handful of officially sanctioned candidates. The few activists who try to effect reform are hobbled by systemic constraints. Protest politics and web-based outrage may send signals to the ruling elites, but these strategies exert no effective pressure. No matter how many people take to the streets in massive marches or in targeted direct actions, they cannot put their hands on the real flows of power, because they do not help produce it. They only consume.

———————

Take the People's Climate March. On a mild, overcast Sunday morning in September 2014, more than three hundred thousand people gathered along the west side of Central Park in New York City for what was billed the "largest climate march in history." The march was being held in advance of a United Nations Climate Summit organized by Secretary General Ban Ki-moon, where more than a hundred heads of state planned to make public commitments to investing in renewable energy, supporting green development in the Global South, developing legally binding regulations on carbon, and working together to help the world's most threatened countries adapt to climate-related risk. I was there as a freelance journalist, to talk to marchers and get a feel for the event, but I was also there as one of the "people" whose march it supposedly was, a concerned global citizen, a worried and guilty American.

I'd marched in New York before, against the war in

Iraq, for Veteran's Day, for gay pride, Halloween, and West Indian Day, and this march like those was a carefully managed event. Indeed, the People's Climate March was a logistical behemoth involving more than 1,500 organizations, each with its own interests and concerns. In order to keep those various organizations united, the march's leading groups, Avaaz and 350.org, decided to forgo message unity, concrete demands, or a clear target: the main message of the march was to be the march itself.

Yet the message the march was supposed to embody was never clear. If the intent was to raise awareness, we would be right to ask what a march is supposed to do that a preponderance of scientific data, decades of research, almost-daily articles in major media outlets worldwide, and nineteen United Nations conferences on the issue since the United Nations Framework Convention on Climate Change first met in Berlin in 1995 could not. If the plan was to convince conservative and mainstream Americans to pay attention to climate change, it's hard to see how a bunch of environmentalists marching through a city most Americans consider a bastion of rich liberals might be expected to accomplish such a thing. If the intent was to demonstrate the political power that the umbrella organizations could motivate in terms of voters, we would be right to ask which voters, where they're from, how they usually vote, and how many there were. None of that information was collected, so one is left to wonder exactly what kind of pressure this march was supposed to put on the American political process, or, for that matter, *any* political process. The march was tightly constrained by police barriers, directed through low-traffic streets far from the location of the United Nations it was supposed to influence, and dumped out in the empty blocks along the Hudson River west of midtown. With no closing rally to unify the protesters, the march ended with an incoherent whimper, as thousands of atomized individuals scattered back to their subways, cars, and digitally wired but politically disconnected lives.

In truth, the People's Climate March was little more than an orgy of democratic emotion, an activist-themed street fair, a real-world analogue to Twitter hashtag campaigns: something that gives you a nice feeling, says you belong in a certain group, and is completely divorced from actual legislation and governance.[73] Given the march's tremendous built-in weaknesses, the best we might have hoped for is that it accomplished nothing. What's more likely, though, is that it siphoned off organizing energy that could have been more useful elsewhere, made a public display of climate activism's political impotence, and soothed hundreds of thousands of people with a false sense of hope.

At the UN Climate Summit two days after the march, attendees didn't even get that spurious elixir; the only cure on offer there was bureaucratic anesthetic. I made it to the upper balcony of the General Assembly Hall just in time to catch the last few minutes of Leonardo DiCaprio's mellifluous opening speech, and sat through much of the day with other journalists and observers in the nosebleed seats, watching the statesmen's dull pageant take place far below. While a range of development and finance schemes were presented as solutions, what the summit came down to was a bleak ritual of stalemate, as if the world's leaders had been cast in a business-class version of Samuel Beckett's *Endgame*.

One after another they stood at the podium mouthing vacuities, boldly committing to toothless, voluntary emissions reductions that are too little, too late. The heads of Russia, Australia, Germany, India, and China didn't even bother showing up. Barack Obama delivered a stirring oration about how everyone needs to work together, chiding China while glorifying America's negligible emissions cuts, ignoring the roles that American coal and American consumption play in stoking China's growth. The president of Tanzania, Jakaya Kikwete, complained about how little African countries have contributed to the problem and how much they stand to suffer, speaking poignantly about Mount Kilimanjaro's disappearing glaciers. Baron Waqa, president of Nauru and chair of the Alliance of Small Island States, argued passionately for more investment to help existentially threatened nations like his own adapt. Chinese Vice Premier Zhang Gaoli addressed the United Nations with a stiff and defensive speech in which he promised that China would peak its carbon emissions "as soon as possible" and reduce its per-capita emissions whenever it got around to it (a position only notionally mitigated by the subsequent US-China climate agreement).

Between marching in the protest on Sunday and going to the UN summit on Tuesday, I rushed up and down Manhattan all day Monday, shuttling between two other events that seemed to promise more concrete results: down at the island's narrow southern tip, a few hundred hard-core activists put their bodies on the line to "Flood Wall Street," while at the Harvard Club in midtown a group of government administrators, corporate representatives, bankers, and economists met to work out a global carbon pricing mechanism at the International Emissions Trading Association.

At Flood Wall Street, which began with a marching band, dancing, and speeches from Naomi Klein, Chris Hedges, Mamadou Goïta, Elisa Estronioli, and others, the message, mission, and enemy were all clearly identified. The protesters were marching to Wall Street in a nonviolent direct action against what they saw as the real problem behind climate change: capitalism. Signs named the usual suspects: the villainous Koch brothers, ExxonMobil, and Shell. The protest's parade of anarchists, environmentalists, and Occupy veterans blocked traffic around Arturo di Modica's "Charging Bull" for a few hours, then made their way to the intersection of Broadway and Wall Street. The New York City Police Department maintained overt control of the situation the entire time, surrounding the marchers with a loose cordon from beginning to end. Helicopters buzzed low overhead, while a nattily dressed community liaison detective observed the situation up close.[74] Wall Street itself had been blocked off by barriers and horse-mounted police long before the protest began; the marchers never stood a chance of getting through. In terms of its own expressed goal, the action was a failure before it even began. The protesters didn't give up, though: they held the intersection of Broadway and Wall, under the steeple of Trinity Church, all

afternoon long. Once the sun set, the cops closed in, order-
ing the protesters to disperse. One hundred and four stal-
wart souls refused to leave and were arrested, including one
sweaty activist in a polar bear suit.

While the rabble faced off against the fuzz downtown,
power-industry elites at the Harvard Club discussed how
to profitably trade waste. On the International Emissions
Trading Association's (IETA) first panel, government representatives talked about how carbon-pricing schemes were
working to reduce carbon emissions in California and Europe, and Tang Jie, the vice mayor of Shenzhen, China,
spoke about his country's carbon market and Shenzhen's
long-term development plans.[75] In contrast to Zhang Gaoli's
stiff, almost hostile speech at the United Nations the next
day, Tang Jie was self-deprecating, effusive, and genial. But
while the body language and tone of Tang Jie's speech was
as obsequious as Zhang Gaoli's was belligerent, the semantic
content of the two speeches was the same: China means to
be rich and it's not going to slow down for anyone.

Following Tang Jie, Harvard economist Robert Stavins
delivered a brief executive summary of the carbon-pricing
study he'd headed, "Facilitating Linkage of Heterogeneous
Regional, National, and Sub-National Climate Policies
Through a Future International Agreement."[76] According
to Stavins, the best result we could expect from the Paris
2015 United Nations Framework Convention on Climate
Change meeting would be a carbon-market deal loosely
linking various national, regional, local, and corporate
systems of carbon pricing and regulation through a uni-

fied monitoring system in which nations would "specify their own targets, actions, policies—or some combination of these—to reduce greenhouse-gas emissions."[77] This sounded ominously vague; I was soon to understand how empty it really was. The last panel featured spokespeople from Statoil, the Environmental Defense Fund, GDF Suez, and Barclay's, all offering more or less hopeful perspectives on the wondrous possibilities of carbon trading, but the key message seemed to be the one delivered by David Hone, Chief Climate Change Advisor for Shell, who translated Stavins's bureaucratese into the much franker point that the only kind of carbon-pricing agreement that had any chance of being signed in Paris would have to be totally nonbinding—which means totally unenforceable.

As with the protesters downtown, everyone in the room seemed to agree that *somebody* needed to do *something* about climate change, and soon. But unlike the protesters at Flood Wall Street who saw capitalism as the underlying problem, the folks at the Harvard Club saw capitalism as the only possible framework for solutions. For the IETA, international legal agreements, corporate management, and market-based mechanisms were the only functional machinery for forging a global response to a global crisis. If this machinery was limited, slow, and troublesome, that didn't mean that it was part of the problem; rather, such limitations were taken as the conditions through which concerned people had to work. Admitting failure wasn't an option, even if the grandstanding at the United Nations and technocratic equivocation at the IETA offered direct evidence for what Thomas

Schelling argues: that the likelihood of the planet's almost two hundred nations drafting a binding emissions-reduction plan is practically nonexistent. The only good odds are for more of the same, which means at best a voluntary and therefore functionally meaningless agreement. Meanwhile, on the very same day that the People's Climate March kicked off, the Global Carbon Project released a new report showing that global greenhouse gas emissions rose 2.3 percent in 2013, with China's emissions growing 4.2 percent, India's growing 5.1 percent, and America's growing 2.9 percent.[78] So much for Obama's oratory.

This seems to be the situation we're stuck in. On the left and on the right, among diplomats, energy company executives, investors, scientists, anarchists, clergy, and activists, serious people are worried about global warming and feel the urgent need to do something about it. Across the spectrum, however, nobody seems to have the tools, clout, or conceptual framework we need to fix it, or even to come up with a good plan to protect ourselves from the greatest dangers. There's no "reset" button for civilization, and no viable plan for transforming global infrastructure, agriculture, and energy networks in the next ten to twenty years. And while smart, dedicated, and thoughtful people fumble with political machinery that doesn't work, such as carbon-pricing markets, protests, and the United Nations, all of us in the Global North go about our business, driving, flying, leaving lights on, running heaters and air conditioners, eating meat, charging our devices, living unsustainable lives predicated on easy consumption.

In a perverse irony, one of the main things that connected the disparate parties at the United Nations, Flood Wall Street, the IETA, and the People's Climate March was a system of cultural technology that is silently burning up masses of carbon while shunting activist outrage into impotent feedback loops. The most common sight at all the events that week was people on their iPhones and Androids, checking email, tweeting, and taking pictures. The global information and communications ecosystem that they were plugged into is now estimated to use about 10 percent of the world's electricity.[79] That ecosystem relies on coal. Every time you check your email, you're heating up the planet. We do it every day. We can't stop. We won't stop.

The problem with the People's Climate March wasn't really that it lacked a goal, or that it was distracting, superficial, and vacuous. The problem with the United Nations isn't that the politicians there are ignorant, hidebound, self-interested, or corrupt. The problem with our response to climate change isn't a problem with passing the right laws or finding the right price for carbon or changing people's minds or raising awareness. *Everybody already knows*. The problem is that the problem is too big. The problem is that different people want different things. The problem is that nobody has real answers. The problem is that the problem is us.

FOUR

THE COMPULSION OF STRIFE

doves exist, dreamers, and dolls;
killers exist and doves, and doves;
haze, dioxin, and days; days
exist, days and death; and poems
exist; poems, days, death

—Inger Christensen, *alphabet*

In the spring of 1931, a Kentucky woman named Florence Reece started rewriting a song she'd written almost twenty years before, when she was twelve years old. The song was originally about a strike her papa, a coal miner, had been marching in, and it was unabashedly polemical. Its title and refrain demand to know: "Which Side Are You On?" A new strike had broken out and Reece was revising her song to fit a new battle, only this time it was her husband who was marching and fighting.

"Which Side Are You On?" went on to become a classic union anthem. Alan Lomax recorded Reece herself singing it in 1937. Years later, it was adapted by the Student Nonviolent Coordinating Committee Freedom Singers as a Civil Rights song. In Reece's 1931 version some of the verses run:

> They say in Harlan County,
> There are no neutrals there.
> You'll either be a union man
> Or a thug for J.H. Blair.
>
> Oh, workers can you stand it?
> Oh, tell me how you can.
> Will you be a lousy scab
> Or will you be a man?
>
> Don't scab for the bosses,
> Don't listen to their lies.
> Us poor folks haven't got a chance
> Unless we organize.

The labor dispute Reece's husband was embroiled in, referred to in the papers of the day as the Harlan County War, was a series of conflicts between Kentucky coal miners and the Harlan County Coal Operators Association. The trouble began when the Association cut the market price of their coal and paid for the cut by taking it out of their miners' wages. Already aggrieved because of predatory lending practices, cheating at the scales, excessive deductions for sick days, and a lack of health benefits, the miners went on strike. The coal companies hired private guards to protect the strikebreaking replacement workers they brought in, their "scabs," and over the next several weeks increasingly violent skirmishes broke out between striking miners and company heavies. These hired guns, the "thugs" of Reece's song, had been deputized by Harlan County sheriff J.H. Blair, and they were given free rein to harass and arrest the miners. Several people were killed on both sides, and after miners ambushed a convoy of guards and killed three of them, the Kentucky National Guard was called in. With the National Guard behind him, Sheriff Blair outlawed all public assemblies, tear-gassed the miners, and within a few weeks broke the strike.

Between 1877 and 1937 there were at least 30 major armed labor conflicts in the United States, including Haymarket, the Pullman Strike, the Lattimer Massacre, the Ludlow Massacre, the Everett Massacre, and the Battle of Matewan. These conflicts were typically between striking workers and motley forces brought in by company bosses, usually including deputized hooligans, scabs, Pinkerton agents, sheriff's posses, and sometimes National Guard troops. In the Ludlow Massacre of 1914, for example, coal company goons backed up by the Colorado National Guard raided a tent colony inhabited by 1,200 mine workers and their families, killing at least 19 people, including several women and children.

The labor movement in those years was an armed fight against corporate tyranny and government repression, a fight for wages, hours, and conditions but also a fight for justice, democracy, and a more equitable form of collective life. Over time and after considerable bloodshed, and after the Great Depression had sent hundreds of thousands of unemployed workers into the streets, the United States government passed the National Labor Relations Act of 1935, enacted various other labor laws, and deployed an array of New Deal social services, all of which helped make life better for working people. The ruling oligarchy did not give workers rights: organized labor won those rights through decades of violent struggle.

The Civil Rights movement, often held up as the great American example of what mobilized masses can do through nonviolent civil disobedience, was, like the labor struggle, a violent, decades-long conflict over social and political power. As journalist and historian Charles Cobb shows in his book, *This Nonviolent Stuff'll Get You Killed: How Guns Made the Civil Rights Movement Possible*, armed resistance and armed defense were foundational to the Civil Rights movement, protecting leaders and organizers, demonstrating political will on the part of black communities and activists, and making nonviolent civil disobedience practicable in a hostile environment.[80]

In 1957, Robert F. Williams became one of the great examples of the effectiveness of armed resistance when he organized the Monroe, North Carolina, chapter of the National Association for the Advancement of Colored People (NAACP) to physically defend the home of his friend and ally Dr. Albert E. Perry. Perry had been targeted by the Monroe Ku Klux Klan (KKK) for leading a protest against rules prohibiting blacks from using a municipal swimming pool. When an armed Klan motorcade came after Perry in his neighborhood, intending to terrorize him into submission, Williams, a US Marine veteran of World War II, had his NAACP chapter meet the Klan with "disciplined, withering volleys" of rifle fire.[81] The Klansmen fled, and the very next day, the Monroe city council banned KKK parades.[82]

Given the intense hostility and violence the Civil Rights movement faced in the South, it's doubtful whether lunch-counter sit-ins, Freedom Rides, and marches would have even been possible without the substantiated threat of armed militancy looming in the background. As James Baldwin wrote in 1963, "Black has *become* a beautiful color—not because it is loved but because it is feared."[83] What's more, the success of the Civil Rights movement can't be properly understood without taking into account federal military interventions, most notably President Eisenhower's deployment of elements of the 101st Airborne Division in 1957 to enforce desegregation in Little Rock, Arkansas; Robert Kennedy's deployment of a contingent of Federal Marshals to protect James Meredith and enforce desegregation at the University of Mississippi in 1962; President John F. Kennedy's subsequent deployment of US Army troops to quell the riots at the University of Mississippi and his deployment of US troops in 1963 to the University of Alabama; and President Johnson's deployment of US troops, federalized National Guards, and Federal Marshals to Alabama in 1965 to protect marchers going to Montgomery from Selma.

The coal miners struggling for a democratic stake in production didn't just protest, share news stories, and post messages. They didn't just march. The African-American activists struggling for civil rights didn't just tweet hashtag campaigns. They didn't just hold meetings. They fought and bled and died for a world they believed in, for a share in the power they produced. Coal strikes throughout the 19th and early 20th century were scenes of violence and turmoil, and it was only through labor's aggressive war on the owners of capital that they were able to win rights that for some decades we considered basic, such as the right to organize or the right to an eight-hour workday. Similarly, the fight to win fundamental civil rights and political equality from segregationists and racists was grinding, dangerous, and aggressive: it strove to take something from Southern white racists that they didn't want to give up, namely power.

Some like to say that "violence never solved anything," but this is a comforting lie, and it's comforting to precisely the wrong people. The real reason that non-violence is considered to be a virtue in the powerless is that the powerful do not want to see their lives or property threatened. As a matter of fact, violence has solved many conflicts. Violence defeated fascism and Nazism in World War II. Violence enforced the end of slavery during the Reconstruction following the American Civil War. Violence freed the American colonies from British rule, just as it freed numerous other colonies across the world from imperial domination. Violence deposed the malignant French aristocracy in 1789 and overthrew the despotic Russian aristocracy in 1917. Violence was central to the successes of the labor movement, and the threat of violence was key to the struggle for Civil Rights. Violence has also been used to conquer vulnerable nations, oppress the weak, torture innocents, threaten critics, force women to submit to rape, pillage cities, and eradicate entire populations. A sword is a sword, whichever way it cuts.

For most of human history, violence has been a central element of social conflict. The first clear evidence of mass human violence is as old as civilization; the first evidence of its end has yet to be seen.[84] According to bioarchaeologist Philip Walker, "As far as we know, there are no forms of social organization, modes of production, or environmental settings that remain free from interpersonal violence for long."[85] As Freud wrote in his famous debate with Einstein on the question of war, "It is a general principle . . . that conflicts of interest between men are settled by the use of violence. This is true of the whole animal kingdom, from which men have no business to exclude themselves."[86] The long record of human brutality seems to offer conclusive evidence that both individual and socially organized violence are as biologically a part of human life as are sex, language, and eating, that aggression and the drive for dominance are neither vestigial atavisms nor social maladaptations but rather species traits, and that we have little reason to hope that war and murder might someday disappear.

Our future promises to be as savage as our past. While it may be true that two centuries of carbon-fueled economic plenty and the widespread ramification of state control have led to a general decrease in daily violence, at least in industrialized nations, the human animal has not purged itself of bloodlust, nor have we put war and violence aside as solutions to our problems. In order to come to an understanding of what it will mean to live and die in the Anthropocene, we must begin to understand what the Greek philosopher Heraclitus meant when he wrote: "It should be understood that war is the common condition, that strife is justice, and that all things come to pass through the compulsion of strife."[87]

When it comes to global warming, differing visions of the human future are already hardening into conflict. Coal and oil companies and their government proxies have made their willingness to use military force to defend themselves and advance their interests spectacularly obvious. The labor wars of the 19th and 20th centuries show this clearly. The brutal decades-long war waged by the Nigerian government against its own people, undertaken with the outright support of Shell and Chevron, is another example, well documented in books such as *A Year and a Day* and *Genocide in Nigeria* by Ken Saro-Wiwa, who was executed for his activism.

You've heard the call: We have to do something. We need to fight. We need to identify the enemy and go after them. Some respond, march, and chant. Some look away, deny what's happening, and search out escape routes into imaginary tomorrows: a life off the grid, space colonies, immortality in paradise, explicit denial, or consumer satiety in a wireless, robot-staffed, 3D-printed techno-utopia. Meanwhile, the rich take shelter in their fortresses, trusting to their air conditioning, private schools, and well-paid guards. Fight. Flight. Flight. Fight. The threat of death activates our deepest animal drives.

The aggression and fear that arise in response to perceived threats are some of the most intense emotions we ever experience. For human society to function at all, these instinctive reactions have to be carefully managed and channeled. Outbreaks of panic and hate are dangerous, but lower levels of aggression and fear help keep a population controllable and productive. Restrained aggression keeps people suspicious of collective action and working hard to overcome their fellows, while constant, generalized anxiety keeps people servile, unwilling to take risks, and yearning for comfort from whatever quarter, whether the dulling sameness of herd thought or the dumb security of consumer goods.

Since at least September 11, 2001, people in the United States have been subject to an unprecedented terror campaign—not from Al Qaeda, but from the United States government. National domestic policy transformed "security" into constant fear, threatening its citizens at every turn: first with alarms of explosions and anthrax, then with prison, austerity-produced structural unemployment, and harassment, and finally with torture, SWAT tanks, snipers, drones, and total surveillance. Owing to the racial logic of US politics, in which white/black is the definitive semiotic distinction structuring American society, most of the government's violence against its own citizens is directed against those with darker skin, but in subtler ways its terror campaign targets every single person who flies coach, watches the news, or uses the Internet.

Fear comes to us every day in our encounters with increasingly militarized police and our humiliating interactions at metal detectors and body-scan machines. Fear comes to us in the absence of job security, in our want of appeal when confronted by institutionalized inequality, and in our mistrust of corrupt institutions. Fear comes to us in widespread surveillance, in the form of a homeless woman or a hospitalized friend without adequate financial support, and in the constant nagging worry that we're not working hard enough, not happy enough, never going to "make it." Fear comes to us in weather porn, unpredictable shifts in formerly stable climate dynamics, and massive storms.

More than in any other way, fear comes to us in images and messages, as social media vibrations, products of cultural technologies that we have interpolated into our lives. Going about our daily business, we receive constant messages of apprehension and danger, ubiquitous warnings, insistent needling jabs to the deep lizard brain. Somebody died. Something blew up. Something might blow up. Somebody attacked somebody. Somebody killed somebody. Guns. Crime. Immigrants. Terrorists. Arabs. Mexicans. White supremacists. Killer cops. Demonic thugs. Rape. Murder. Global warming. Ebola. ISIS. Death. Death. Death.

Sociologist Tom Pyszczynski writes: "People will do almost anything to avoid being afraid. When, despite the best efforts, [fear and anxiety] do break through, people go to incredible lengths to shut them down."[88] Sometimes when these vibrations shake us, we discharge them by passing them on, retweeting the story, reposting the video, hoping that others will validate our reaction, thus assuaging our fear by assuring ourselves that collective attention has been alerted to the threat. Other times we react with aversion, working to dampen the vibrations by searching out positive reinforcements, pleasurable images and videos, something funny, something—anything—to ease the fear. We buy something. We eat food. We pop a pill. We fuck.

In either passing on the vibration or reacting against it, we let the fear short circuit our own autonomous desires, diverting us from our goals and loading ever more emotional static into our daily cognitive processing. We become increasingly distracted from our ambitions and increasingly susceptible to such distraction. And whether we retransmit or react, we reinforce channels of thought, perception, behavior, and emotion that, over time, come to shape our habits and our personality. As we train ourselves to resonate fear and aggression, we reinforce patterns of thought and feeling that shape a society that breeds the same.

Fight-or-flight is compelling because it serves essential evolutionary purposes. It increases alertness and adrenaline flow, and generally works to keep the human animal alive. As we proceed into the Anthropocene, though, capitalism's cultural machinery for balancing fear and aggression against desire and pleasure is grinding and sputtering sparks. What cultural theorist Lauren Berlant has identified as the "cruel optimism" of a system sustained by hopes that can never be fulfilled mixes dangerously with an atmosphere of beleaguered anxiety, increasing frustration with working-class and middle-class economic stagnation, and a pervasive sadistic voyeurism that grows by what it feeds on.[89] While America's fraying social infrastructure holds together, our fear and aggression can be channeled into labor, consumption, and economic competition, with professional sports, hyperviolent television, and occasional protests to let off steam. Once the social fabric begins to tear, though, we risk unleashing not only rioting, rebellion, and civil war, but homicidal politics the likes of which should make our blood run cold.

———

Consider: Once among the most modern, Westernized nations in the Middle East, with a robust, highly educated middle class, Iraq has been blighted for decades by imperialist aggression, criminal gangs, interference in its domestic politics, economic liberalization, and sectarian feuding. Today it is being torn apart between a corrupt petrocracy, a breakaway Kurdish enclave, and a self-declared Islamic fundamentalist caliphate, while a civil war in neighboring Syria spills across its borders. These conflicts have likely been caused in part and exacerbated by the worst drought the Middle East has seen in modern history. Since 2006, Syria has been suffering crippling water shortages that have, in some areas, caused 75 percent crop failure and wiped out 85 percent of livestock, left more than 800,000 Syrians without a livelihood, and sent hundreds of thousands of impoverished young men streaming into Syria's cities.[90] This drought is part of long-term warming and drying trends that are transforming the Middle East.[91] Not just water but oil, too, is elemental to these conflicts. Iraq sits on the fifth-largest proven oil reserves in the world. Meanwhile, the Islamic State has been able to survive only because it has taken control of most of Syria's oil and gas production. We tend to think of climate change and violent religious fundamentalism as isolated phenomena, but as Retired Navy Rear Admiral David Titley argues, "you can draw a very credible climate connection to this disaster we call ISIS right now."[92]

A few hundred miles away, Israeli soldiers spent the summer of 2014 killing Palestinians in Gaza. Israel has also been suffering drought, while Gaza has been in the midst of a critical water crisis exacerbated by Israel's military aggression. The International Committee for the Red Cross reported that during summer 2014, Israeli bombers targeted Palestinian wells and water infrastructure.[93] It's not water and oil this time, but water and gas: some observers argue that Israel's "Operation Protective Edge" was intended to establish firmer control over the massive Leviathan natural gas field, discovered off the coast of Gaza in the eastern Mediterranean in 2010.[94]

Meanwhile, thousands of miles to the north, Russian-backed separatists fought fascist paramilitary forces defending the elected government of Ukraine, which was also suffering drought.[95] Russia's role as an oil and gas exporter in the region and the natural gas pipelines running through Ukraine from Russia to Europe cannot but be key issues in the conflict. Elsewhere, droughts in 2014 sent refugees from

Guatemala and Honduras north to the US border, devastated crops in California and Australia, and threatened millions of lives in Eritrea, Somalia, Ethiopia, Sudan, Uganda, Afghanistan, India, Morocco, Pakistan, and parts of China. Across the world, massive protests and riots have swept Bosnia and Herzegovina, Venezuela, Brazil, Turkey, Egypt, and Thailand, while conflicts rage on in Colombia, Libya, the Central African Republic, Sudan, Nigeria, Yemen, and India. And while the world burns, the United States has been playing chicken with Russia over control of Eastern Europe and the melting Arctic, and with China over control of Southeast Asia and the South China Sea, threatening global war on a scale not seen in seventy years. This is our present and future: droughts and hurricanes, refugees and border guards, war for oil, water, gas, and food.

We experience this world of strife today in one of two modes: either it is our environment, and we are in it, or it comes to us as images, social excitation, retransmitted fear. People are fighting and dying in ruined cities all over the planet. Neighbors are killing each other. Old women are bleeding to death in bombed rubble and children are being murdered, probably as you read this sentence. To live in that world is horrific. Constant danger strains every nerve. The only things that matter are survival, killing the enemy, reputation, and having a safe place to sleep. The experience of being human narrows to a cutting edge.

I remember living in that world many years ago in occupied Baghdad. Today that world seems impossibly distant, yet every day it presses in on me in a never-ending stream of words, images, appeals, and reports. I see videos. I read stories. I see pictures of this or that suffering or injustice and I am moved. To act, perhaps, but more accurately to emote. To react. To feel. To perform. We do not usually ask where these feelings come from or who they serve, but we all know that the cultural technologies transmitting these affective vibrations are not neutral: news outlets shape information to fit their owners' prejudices, while Facebook, Twitter, and Google shape our perceptions through hidden algorithms. The specialization and demographic targeting of contemporary media tend to narrow the channels of perception to the point that we receive only those images and vibrations which already harmonize with our own prejudices, our own pre-existing desires, thus intensifying our particular emotional reactions along an increasingly limited band, impelling us to discharge our emotions within the same field of ready listeners, for which we are rewarded with "Likes" and "Favorites." Our consciousness is shaped daily through feedback systems where some post or headline provokes a feeling and we discharge that feeling by provoking it in others. Social media like Facebook crowdsource catharsis, creating self-contained wave pools of aggression and fear, pity and terror, stagnant flows that go nowhere and do nothing.

Pictures of children killed by bombs or police, or pictures of the devastation left in the wake of a tropical storm may move me to sadness and horror. Retransmitting such images will pass along that sadness and horror. My act of transmission will mark me as someone who has feelings about these things and who condemns them. I can rationalize my retransmission by saying that I am "raising aware-ness" or trying to influence public policy: I want my fellow citizens to be as horrified as I am, so they'll think like I do, or so they'll vote for a representative who works to prevent such horrors from happening, or maybe so that if enough of us all think the same way and feel the same way, the organs and institutions of power will be forced to hear us and align themselves along our vibrations, the way a honeybee colony will pick a site for a new hive through the dance of its ad-vance guard scouts.

These are perfectly reasonable human assumptions, be-cause that is how physical human collectives function. Any-one who has been in a crowd, a basketball team, a nightclub, a choir, or a protest knows how bodies resonate together. But politics is the energetic distribution of bodies *in systems*, and we live in a system of carbon-fueled capitalism that we shouldn't expect to work in physical human ways for sev-eral reasons, especially when it comes to responding to the threat of global warming. First, our political and social me-dia technologies are not neutral, but have been developed to serve particular interests, most notably targeted advertising, concentration of wealth, and ideological control, and the vibrations that seem to resonate most strongly along these channels are envy, adulation, outrage, fear, hatred, and mind-less pleasure. Second, the more we pass on or react to social vibrations, the more we strengthen our habits of channeling and the less we practice autonomous reflection or indepen-dent critical thought. With every protest chant, retweet, and Facebook post, we become stronger resonators and weaker thinkers. Third, however intense our social vibrations grow, they remain locked within machinery that offers no political leverage: they do not translate into political action, because they do not connect to the flows of power. Finally, while the typical collective human response to threat is to identify an enemy, pick sides, and mobilize to fight, global warming offers no apprehensible foe.

That hasn't stopped people from trying to find one. The Flood Wall Street protesters say the enemy is American corporations. Tanzania's Jakaya Kikwete and Nauru's Baron Waqa say the problem is the United States and Great Britain. Shell Oil and the Environmental Defense Fund seem to think that it's intractable UN bureaucracy that's holding us up. Barack Obama has implied that it's China. Tea Party Republicans would blame Barack Obama, I'm sure, if they admitted that global warming was actually happening and caused by human activity. Meanwhile, NPR-listening liberals want to believe that Tea Party Republicans are responsible, so that they can frame the problem as one amenable to solution by moral education and enlightened consumerism, as if it were all a matter of convincing people to eat more kale and drive electric cars. One climate activist has argued that just 90 companies are responsible for almost two-thirds of all historical greenhouse gas emissions, which conveniently absolves billions of automobile drivers, airline passengers, meat eaters, and cellphone users of responsibility.[96] The enemy isn't *out there* somewhere—the enemy is ourselves. Not as individuals, but as a collective. A system. A hive.

How do we stop ourselves from fulfilling our fates as

suicidally productive drones in a carbon-addicted hive, destroying ourselves in some kind of psychopathic colony collapse disorder? How do we interrupt the perpetual circuits of fear, aggression, crisis, and reaction that continually prod us to ever more intense levels of manic despair? One way we might begin to answer these questions is by considering the problem of global warming in terms of Peter Sloterdijk's idea of the philosopher as an interrupter:

> We live constantly in collective fields of excitation; this cannot be changed so long as we are social beings. The input of stress inevitably enters me; thoughts are not free, each of us can divine them. They come from the newspaper and wind up returning to the newspaper. My sovereignty, if it exists, can only appear by my letting the integrated impulsion die in me or, should this fail, by my retransmitting it in a totally metamorphosed, verified, filtered, or recoded form. It serves nothing to contest it: I am free only to the extent that I interrupt escalations and that I am able to immunize myself against infections of opinion. Precisely this continues to be the philosopher's mission in society, if I may express myself in such pathetic terms. His mission is to show that a subject can be an interrupter, not merely a channel that allows thematic epidemics and waves of excitation to flow through it. The classics express this with the term 'pondering.' With this concept, ethics and ener-

getics enter into contact: as a bearer of a philosophical function, I have neither the right nor the
desire to be either a conductor in a stress-semantic
chain or the automaton of an ethical imperative.[97]

Sloterdijk compares the conception of political function as collective vibration to a philosophical function of interruption. As opposed to disruption, which shocks a system
and breaks wholes into pieces, interruption suspends con-
tinuous processes. It's not smashing, but sitting with. Not
blockage, but reflection.

Sloterdijk sees the role of the philosopher in the human swarm as that of an aberrant anti-drone slow-dancing
to its own rhythm, neither attuned to the collective beat
nor operating mechanically, dogmatically, deontologically,
but continually self-immunizing against the waves of social
energy we live in and amongst by perpetually interrupting
its own connection to collective life. So long as one allows
oneself to be "a conductor in a stress-semantic chain," one
is strengthening channels of retransmission regardless of
content, thickening the reflexive connective tissues of mass
society, making all of us more susceptible to such viral phenomena as nationalism, scapegoating, panic, and war fever.
Interrupting the flows of social production is anarchic and
counterproductive, like all good philosophy: if it works, it
helps us stop and see our world in new ways. If it fails, as it
often and even usually does, the interrupter is integrated,
driven mad, ignored, or destroyed.

What Sloterdijk helps us see is that responding autonomously to social excitation means not reacting to it, not passing it on, but interrupting it, then either letting the excitation die or transforming it completely. Responding freely to constant images of fear and violence, responding freely to the perpetual media circuits of pleasure and terror, responding freely to the ongoing alarms of war, environmental catastrophe, and global destruction demands a reorientation of feeling so that every new impulse is held at a distance until it fades or can be changed. While life beats its red rhythms and human swarms dance to the compulsion of strife, the interrupter practices dying.

FIVE

A NEW ENLIGHTENMENT

Let my eyes see the sun and be sated with light!
The darkness is hidden, how much light is there left?
When may the dead see the rays of the sun?

—*The Epic of Gilgamesh*, Tablet IX[98]

Death begins as soon as we are born. From our first moments in the world, blinking and crying in the light, we fly an unwavering arc to the grave. We are mortal, material, corporeal beings, blessed with quickening but doomed to decay. In a very important sense, we don't need to learn anything at all about dying: it's the one thing in life we can absolutely count on getting done.

But while dying may be the easiest thing in the world to do, it's the hardest thing in the world to do well—we are predisposed to avoid, ignore, flee, and fight it till the very last hour. We are impelled in our deepest being to struggle against it. Every time you feel hunger or taste ambition, every time your body tingles with lust or your heart yearns for recognition, every time you shake with anger or tremble in fear, that's the animal in you striving for life. We fall into the world caught between two necessities, compelled to live, born to die, and reconciling them has forever been one of our most challenging puzzles. The pieces just don't fit together.

Much of our energy is spent in denial. Some argue that denying the fact of death is the root and germ of human culture itself, from our first burial mounds and ancestor-worship to plastic surgery and the space program.[99] "The idea of death, the fear of it, haunts the human animal like nothing else," wrote anthropologist Ernest Becker in his Pulitzer Prize–winning book *The Denial of Death*. "It is a mainspring of human activity—activity designed largely to avoid the fatality of death, to overcome it by denying in some way that it is the final destiny for man."[100] Throughout human history, we have invested innumerable hours of labor in countless luminous visions of the afterlife, both physical and metaphysical: Heaven and Hell, family and nation, capitalism and *Star Trek*. We have children and pressure our children to give us grandchildren, sending our genes into the future. We build pyramids, cathedrals, temples, mosques, monuments, and skyscrapers to prove to ourselves that some part of us will survive beyond our own end. And when our buildings crumble and our gods grow weak, we distract ourselves with pleasure or rally ourselves to war. As men who have experienced war have testified since the days of Homer, there is "a joyous feeling in the safety of killing" that washes away the bitter taste of death.[101]

Accepting the truth of our end is the beginning of wisdom. When Montaigne wrote that "To philosophize is to learn how to die," he was working in and with a philosophical tradition that was already centuries old. Citing one of his key predecessors, the Roman orator Cicero, Montaigne wrote: "Cicero says that to philosophize is nothing else but to prepare for death. This is because study and contemplation draw our soul out of us to some extent and keep it busy outside the body; which is a sort of apprenticeship and semblance of death. Or else it is because all the wisdom and reasoning in this world boils down finally to this point: to teach us not to be afraid to die."[102] Cicero was in his turn reworking Plato's account of the death of Socrates in the *Phaedo*, where Socrates argues that philosophy is the practice of learning how to separate the soul from the body.[103]

Philosophical humanism in its most radical practice is the disciplined interruption of somatic and social flows, the detachment of consciousness from impulse, and the condensation of conceptual truths out of the granular data of experience. It is the study of "dying and being dead," a divestment from *this* life in favor of deeper investments in a life beyond ourselves. In recognizing the dominion of death and the transience of individual existences, we affirm a web of being that connects past to future, them to us, me to you. "One is responsible to life," wrote James Baldwin. "It is the small beacon in that terrifying darkness from which we come and to which we return. One must negotiate this passage as nobly as possible, for the sake of those who are coming after us."[104]

Learning to die is hard. It takes practice. There is no

royal road, no first-class lane. Learning to die demands daily cultivation of detachment and daily reminders of mortality. It requires long communion with the dead. And since we can't ever really know how to do something until we do it, learning to die also means accepting the impossibility of achieving that knowledge as long as we live. We will always be practicing, failing, trying again and failing again, until our final day. Yet the practice itself is the wisdom. In the words of Zen master Dōgen: "To practice the way single-heartedly is, in itself, Enlightenment."[105]

As I learned in Iraq and have had to learn again and again, the practice of learning to die is the practice of learning to let go: Learning to die means learning to let go of the ego, the idea of the self, the future, certainty, attachment, the pursuit of pleasure, permanence, and stability. Learning to let go of salvation. Learning to let go of hope. Learning to let go of death. It means realizing with the Stoic philosopher Marcus Aurelius that

> Of human life the time is a point, and the substance is in a flux, and the perception dull, and the composition of the whole body subject to putrefaction, and the soul a whirl, and fortune hard to divine, and fame a thing devoid of judgment. And, to say all in a word, everything which belongs to the body is a stream, and what belongs to the soul is a dream and vapor, and life is a warfare and a stranger's sojourn, and after-fame is oblivion.[106]

Learning to die means realizing along with the German philosopher G.W.F. Hegel that human consciousness operates through a dialectic of negation, and that enlightened self-consciousness is consciousness of one's own limits—and of one's own death:

> The human being is this Night, this empty nothing which contains everything in its simplicity—a wealth of infinitely many representations, images, none of which occur to it directly, and none of which are not present. This is the Night, the interior of human nature, existing here—*pure Self*—in phantasmagoric representations it is night everywhere: here a bloody head suddenly shoots up and there another white shape, only to disappear as suddenly. We see this Night when we look a human being in the eye, looking into a Night which turns terrifying. For from his eyes the night of the world hangs out toward us.[107]

This Night was the face I saw when I confronted the fact of my own mortality. It's the face we all see sooner or later, because it's our own face—our own consciousness, our own death mask. It waits for us in the mirror.

Accepting this emptiness, letting go of my self, was only the first step in coming to understand my responsibility to and participation in a larger collective self, a kind of human existence transcending any particular place or time, going back to our first moments in Africa 200,000 years ago, and living on in the dim, fraught future of the Anthropocene. We are humanity. We are the dead. They have become us, as we will become the dead of future generations.

We are born half-blind, confused, wired into a world we don't understand. Within the night of this world, we apprehend our future as a field of freedom. We face this freedom as individuals, fully in the present, yet our actions are determined by the past and take on their full meaning only in the future. As we gain in wisdom, individual consciousness reveals its complex entanglements with collective life, history, and the universe. We live in and orient our existence through conceptual and narrative structures that rationalize our impulses, pattern our habits, and connect our behaviors to collective rhythms. These conceptual and narrative structures are the cultural technology through which we make meaning and shape our desires. Facebook shapes desire differently than does the Koran, each of which shapes desire differently than does a West Elm catalog, Emily Dickinson's poetry, Tài Chi, the Igbo New Yam festival, democracy, Passover, "the market," or Mexican telenovelas.

The only inherent trait of the human ape that differentiates us from other animals is our knack for collective symbolic manipulation. Other species besides Homo sapiens communicate with language, organize socially, build structures, use tools, laugh, and show emotions. Even fire and simple social technologies were part of an inheritance passed down from Homo erectus and the Neanderthal.[108] Sometime in the icy depths of prehistory, though, our species began developing advanced symbolic communication beyond anything that had ever been seen before. We learned how to make the dead speak, and to speak ourselves to the yet unborn. We learned to see into the future. We learned to abstract from the present a conceptual reality transcending time and space. Through the ice ages of the past and into the long summer of the Holocene we carried tools, furs, fire, and our greatest treasure and most potent adaptive technology, the only thing that might save us in the Anthropocene, because it is the only thing that can save those who are already dead: memory.

———————

Slightly more than 3,000 years ago, a band of Mycenaean shepherd-warriors raided and burned a walled city on the Anatolian coast. What exactly happened remains a matter of conjecture, since the empirical evidence is sketchy. On the one hand, we have the archaeological remains of a destroyed town, some evidence of fire, and a few arrowheads, all of which was only excavated in the last hundred and thirty years. On the other hand, we have two long poems, probably composed orally and written down about 2,700 years ago, that tell the stories of these shepherd-warriors and their raid. Those two poems, originally recorded on papyrus or parchment rolls, grew to become authoritative texts for the ancient Greek city-states, the Roman empire, Byzantine civilization, and modern European and American literature, helping inspire and influence thinkers from Plato to Milton, Alexander Pope to Thomas Jefferson, Simone Weil to

Derek Walcott. These long poems survived by being passed on from family to family, preserved in temples and libraries, and recopied by monks and poets.

While hundreds of scraps of Homeric poetry date as far back as 2,300 years ago, the oldest surviving full copy of Homer's *Iliad* only dates to the 10th century CE. "The creation of this great book was no routine act of copying but a major scholarly enterprise," writes classicist Martin West.[109] This book, the "Venetus A" manuscript, is stored in the Public Library of St. Mark in Venice, and has been digitized by Harvard University.[110] You can look at it online.[111]

About 400 years after somebody first transcribed the *Iliad*, an Athenian vintner and war veteran named Aeschylus wrote a quartet of plays for performance in the annual religious festivals honoring Dionysus, the god of wine. Three plays in the quartet tell the bloody tale of the House of Atreus and the origins of Athenian law. In the first play, King Agamemnon returns home from the Trojan War and is murdered by his wife Clytaemnestra, in revenge for an even more heinous crime: years ago, Agamemnon had sacrificed their daughter to secure his army's passage to Troy. Agamemnon's war-trophy rape-slave Cassandra prophesied the king's murder, but no one heard her shouts until they turned to screams, as Clytaemnestra struck her down alongside her captor. The queen got her revenge, but in the second play, *The Libation Bearers*, Clytaemnestra is murdered in turn by her son, Orestes. These plays give narrative shape to the compulsion of strife, the constant turn and return of violence, blood guilt and blood vengeance, the endless cycle of fear and aggression, desire and death.

Yet they also show us a way out of that cycle: In the third play, *The Eumenides*, Apollo and Athena argue over Orestes' fate and eventually decide to exonerate him. The Erinyes, the "Furies" who embodied the old law of vengeance and had meant to hound Orestes to death, are transformed by Athena into the Eumenides, the "Kindly Ones," and made the guardians of Athenian law. Through reflection, justice is transformed from revenge into mercy.

The final play of the quartet, the comic satyr play *Proteus*, is lost. Indeed, the three plays I've described are the only extant dramatic trilogy to survive from ancient Greece at all. In total, we have only seven of Aeschylus's plays, out of an estimated seventy to ninety that he wrote. This is fairly representative. The vast majority of classical Greek culture has disappeared. The little that persists, though, including much of Plato and Aristotle, some plays by Sophocles and Euripedes, fragments of poems by Sappho and Callimachus, and the histories of Xenophon, Herodotus, and Thucydides, is an immensity of riches compared to the rumors and scraps that have endured from Homer's time, only a few hundred years before, or from the Mycenaean and Hittite cultures that Homer's poems make legends of.

Why even bother with these relics of a savage time? What do Homer and Aeschylus have to do with ISIS or global warming? When we look at them closely, these ancient Greeks grow to seem strange, even barbaric. They didn't believe in free will like we do, they didn't believe in progress, and they certainly didn't believe in universal human rights. The Greek concept of fate is hard to bear. Their ideas about justice seem cruel: kings and gods are capriciously brutal, and transgression is often punished with awesome suffering. Even in familiar stories such as *Oedipus Rex* and the *Iliad*, these archaic people act in ways that strike us as not just primitive but irrational, superstitious, even insane. Yet they are us: not only historically, but genetically, technologically.

A hairsbreadth of human time separates Aeschylus from the present, and in that moment grows everything we take for modernity. Our language, our thought, our architecture, and our culture carry ancient memories embedded in grammar, vocabularies, and syntax. Our symbolic-epistemological structures of cognition and discourse not only have histories, but are *made* of history, are made from words such as "*astron*" and "*nomos*," "*ge*" and "*logos*," "*anthropos*" and "*kainós.*" Our thoughts and narratives are built from sedimented archetypes such as Achilles and Cassandra, concepts such as fate, justice, and democracy, and the very idea of an idea (*eidos*, meaning that which is seen, form, or shape).

Attending to the historical and philological genealogies of our current conceptual, symbolic structures of existence helps us recognize who we are, who we have been, and who we might become. The comparative study of human cultures across the world and through time helps us see that our particular way of doing things right here, right now, is a contingent adaptation to particular circumstances, yet at the same time an adaptation built with universal human templates of meaning-making and symbolic reasoning, with tools and technologies we have inherited from the past. I've relied mainly on Greek examples, but the roots of our contemporary global civilization are also Akkadian, Sumerian, Chinese, Indian, Mesoamerican, Judaic, Egyptian, Nubian, Thule, Dorset, and Finno-Ugric. Anywhere humans live, we make meaning. The record of that wisdom, the heritage of the dead, is our most valuable gift to the future.

The study of the humanities is nothing less than the patient nurturing of the roots and heirloom varietals of human symbolic life. This nurturing is a practice not strictly of curation, as many seem to think today, but of active attention, cultivation, making and remaking. It is not enough for the archive to be stored, mapped, or digitized. It must be *worked*. As Hannah Arendt writes:

> If it is true that all thought begins with remembrance, it is also true that no remembrance remains secure unless it is condensed and distilled into a framework of conceptual notions within which it can further exercise itself. Experiences and even the stories which grow out of what men do and endure, of happenings and events, sink back into the futility inherent in the living world and the living deed unless they are talked about over and over again. What saves the affairs of mortal men from their inherent futility is nothing but this incessant talk about them, which in its turn remains futile unless certain concepts, certain guideposts for fu-

ture remembrance, and even for sheer reference, arise out of it.[112]

Against the futility of life without memory, we have only this delicate accretion. Papyrus rots, paper burns, museums get sacked, hard drives crash. The fragility of our collective cultural enterprise is well illustrated by the epigraph heading this chapter and the long poem it comes from. The *Epic of Gilgamesh*, one of the oldest works of literature on Earth, was probably first inscribed on clay tablets sometime in the 21st or 20th century BCE, later lost for nearly 2,500 years, then recovered in the middle of the 19th century by an Assyrian archaeologist from Mosul named Hormuzd Rassam.[113] The only reason the *Epic* survived was because it had been copied out by ancient scribes as rote training for more "important" bureaucratic and commercial work.

The *Epic* tells the story of Gilgamesh, king of Uruk, a "wild bull on the rampage" admired for his strength but resented for his despotism.[114] As the *Epic* recounts: "The young men of Uruk he harries without warrant, / Gilgamesh lets no son go free to his father. / By day and by night his tyranny grows harsher." The women of Uruk pray for relief, complaining of Gilgamesh's insistence on taking the virginity of all the brides in his city. The gods hear the women's prayer, and one of them makes a wild man to match Gilgamesh, a hair-covered man-beast named Enkidu, the "off-spring of silence." Sent into the world, Enkidu runs with gazelle herds and sabotages hunters' traps till Gilgamesh hears of his mischief and sends a woman from Uruk named Shamhat to tame him. She finds Enkidu, seduces him, dresses him in clothes, teaches him how to eat bread and drink ale, and tells him of the wonders of city life. When Enkidu hears about Gilgamesh's habit of taking other men's brides on their wedding night, though, "his face pale[s] in anger" and he speeds to Uruk. Once there, Enkidu challenges Gilgamesh, but their fight ends in a draw, with the two men kissing and becoming friends.

In classic buddy movie style, Enkidu and Gilgamesh team up and journey to the Forest of Cedar, "the secret abode of the gods," where they kill the ogre Humbaba and take his majestic trees. When they return to Uruk, the goddess of love, Ishtar, tries to seduce Gilgamesh. He rejects her, so she sends down the monstrous Bull of Heaven to destroy Uruk and kill Gilgamesh. Enkidu and Gilgamesh turn the tables and slaughter the bull instead. That night, Enkidu has a dream in which the gods declare that the heroes have gone too far, and that one of them will have to die. They name Enkidu. He wakes and recounts his dream, then, after much lamenting, succumbs to a fatal illness.

Gilgamesh is overcome with grief. "Afraid of death," he wanders the Earth weeping for his friend, looking for passage to the Netherworld in the hopes of uncovering the secret of immortality. Gilgamesh finds the gate to the Netherworld and passes through twelve "double-hours" of darkness, emerging in a beautiful garden where he meets Uta-napishti, an immortal elder who had survived the legendary Deluge. Gilgamesh demands the secret of immortality, and Uta-napishti tells the story of how before the

Deluge he was ordered by the gods to build a great boat to carry "the beasts of the field, the creatures of the wild, and members of every skill and craft." One day, as the gods foretold, the rain began to fall.

> For a day the gale winds flattened the country,
> quickly they blew, and then came the Deluge.
> Like a battle the cataclysm passed over the
>   people. . . .
> For six days and [seven] nights,
> there blew the wind, the downpour,
> the gale, the Deluge, it flattened the land.

After the rain stopped, Uta-napishti released a dove and a sparrow, both of which came back, and finally a raven, which did not. Once the waters receded, the gods made Uta-napishti and his wife immortal. Uta-napishti tells Gilgamesh that he too can be immortal, if he goes six days and seven nights without sleeping.

Gilgamesh tries hard to say awake, but his eyes close the very first night and he sleeps for a week. When Uta-napishti rouses him, Gilgamesh is so distraught that the old man feels bad for him and gives him a miraculous rejuvenating plant. On his way home, though, Gilgamesh loses the plant to a snake, who steals it from him in the night. The mighty king curses the futility of existence: "For whom, Ur-shanabi, toiled my arms so hard?" he wails. "For whom ran dry the blood of my heart?"

We might answer: For us. For the future. Almost four

thousand years out of the past, the *Epic of Gilgamesh* carries forward a story of civilization. Its first main narrative, the taming of the wild man Enkidu, is reiterated in its final act, in which the rampaging Gilgamesh learns to accept the truth of death. Enkidu and Gilgamesh represent two phases of social development, the hunter-gatherer and the agricultural. The absolutist demands of agricultural civilization, embodied in Gilgamesh, are tyrannical without the tempering wisdom of the wild, but when civilization and wildness join forces, they create an all-consuming war machine that disrupts the sacred order. The gods decree the wild man must die and the mighty king submit to human limits.

The *Epic of Gilgamesh* also offers a lesson in the importance of sustaining and recuperating cultural heritage in the wake of catastrophic climate change. As the *Epic*'s prologue reads, Gilgamesh "saw what was secret, discovered what was hidden, / he brought back a tale of before the Deluge." Translator and scholar Andrew George explains: "The implication of the prologue of the epic is that Gilgamesh played a key role in restoring the antediluvian order after the Flood, particularly in restoring the cults of the gods to their proper glory. . . . It did not suit the poet's need to include more than allusions to it, but evidently Gilgamesh was responsible for re-civilizing his country."[115]

Gilgamesh lives on in death as a legend, a symbol, a reflection on ourselves. In this way, the *Epic* represents not only the fragility of our deep cultural heritage, but its persistence. I saw this firsthand when I talked to the Iraqi heavy metal band Acrassicauda about their first full-length album, which they titled *Gilgamesh*.[116] Marwan Hussein, the band's drummer and guiding force, told me that he'd decided to turn to the epic as a template for their album because it connected back to Iraq's ancient Sumerian roots while at the same time offering a way of understanding the situation there now. "The *Epic of Gilgamesh* is a very weird story," Hussein told me, "and in a way very modern. It's a way to tell the story of what's happening today in the Middle East: Gilgamesh was a tyrant. He was not a very good king. He was weak. The way I see him, he was a lost soul until the very last, when he comes to terms with his mortality. In the end, it's a story about rebirth."

Rebirth is a resonant concept for the members of Acrassicauda: they grew up in Saddam Hussein's Iraq, lived through the US invasion in 2003, escaped the country's brutal sectarian civil war in 2005 and 2006, and, with the help of *VICE* magazine co-founder Suroosh Alvi and a documentary he made about them (*Heavy Metal in Baghdad*), made their way through Syria and Turkey to the United States.[117] Fifteen years after they formed a band in Baghdad, teaching themselves to play Metallica and Slayer off bootleg cassettes, they finally released their first album—an album which is at once a testament to their artistic ambitions, a reflection on their unique historical situation, and a remembrance of their heritage. As Hussein told me, "We tried to make a Middle Eastern metal album. We did the whole thing. We did the percussion, we did darbuka, we did the time signature, we did the singing. We wanted the album to be as Arabic as we could make it."

Acrassicauda are doing the hard work of cultivating and remaking our interwoven cultural technologies. They didn't have to. Their fate could have been completely different. In Iraq, in Syria, even as refugees living in New Jersey, Acrassicauda would have had every excuse to confuse acceptance of their situation with giving up, or to mistake hatred for justice. They might have stayed in Baghdad and fought to free their country from foreign invaders. Many did. Many chose another side, and fought for the Americans, or for Al Qaeda. Instead, Hussein and his bandmates chose music and art. They chose to explore and strengthen the connections between cultures, our shared rhythms and common traditions. They embody for their fans the realized hope of transcending parochialism and strife while staying connected to tradition, the courage of surviving war not by violence but through decency and mutual support, and the lived possibility that we may—like the troubled king Gilgamesh—learn to die and yet be reborn.

More than 6,000 years before Homer sang of Achilles' rage and more than 5,000 years before the *Epic of Gilgamesh* was written, humans living in what is now China, Serbia, and Iraq began to make marks on bone and clay. Agricultural and astronomical knowledge, relatively stable, had been stored in memorized rhythmic speech, but over time the novel technology of writing allowed us to keep better track of more changeable data: property, grain stores, trade. The practice of writing integrated older social technologies developed for other purposes, and as writing grew in sophistication and importance it absorbed yet more. Ritual song, image-making, mythology, religion, rhymed speech, memorization, metonymic association, metaphoric abstraction—technologies of social attunement, information storage, ideological mapping, emotional regulation, and political organization—were all within a few thousand years intimately wired into graphic representations of language.

First through clay and stone, then papyrus, vellum, and paper, writing became the single most important human development after agriculture and before the steam engine, leading to widespread social transformations and enabling the creation of incredible new technologies. From writing, the ancient Sumerians in Uruk developed mathematics, allowing them to abstractly yet accurately model physical relations in the world. Written laws gave concrete form to political and religious authority. Being able to track and compare information allowed observers to adjudicate evidence, which gave rise over centuries to empiricism and the scientific method. The development of the printing press in 1450 and the later proliferation of affordable books (by means of hand presses in the 18th century, then steam and coal-powered industrial presses in the 19th and 20th centuries) meant that vast amounts of stored information could be accessed by anyone able to read. The fullest expression of human life soon came to be seen as synonymous with literacy itself. With the advent of mass-produced sound and image reproduction technologies in the 19th and early

20th centuries (phonograph, radio, film, TV), humanism-as-literacy was superseded, but with the rise of personal computers and the Internet, it has been re-integrated and transformed into humanism-as-digital-literacy, or what we might call photohumanism.

A new form of life has become evident: humanity has revealed itself as collective energy, light swarming across a darkened planet, a geological forcing, data and flow. We live in networks, webs, and hives, jacked in to remote-controlled devices and autonomous apps, moments of being in time, out of time. No longer individual subjects or discrete objects, we have become vibrations, channelers, tweeters and followers. By connecting us through our devices, photohumanist technologies enable collective wiring at tremendous speeds, even faster than those offered by radio, print, television, and film.

Just as those older technologies offered both great potential and great danger, so too do the newer technologies of photohumanism. *Homo sapiens*—perhaps now *Homo lux*—remains biologically reactive, easily panicked, all too quickly stirred to hatred. Jaron Lanier warned in 2006 of what he called "digital Maoism": "It is at least possible that in the fairly near future enough communication and education will take place through anonymous Internet aggregation that we could become vulnerable to a sudden dangerous empowering of the hive mind. History has shown us again and again that a hive mind is a cruel idiot when it runs on autopilot."[118] The dangers of collective madness, witch hunts, and totalitarian war exist in new social networks as much as or even more than they did in early 20th-century radio broadcasts, newspaper stories, and cinema reels. Our collective obligation to maintain traditional humanistic study in the photo-humanist era is at once developmental and prophylactic: we must practice interruption to nurture new flows and at the same time to guard against them.

As we struggle, awash in social vibrations of fear and aggression, to face the catastrophic self-destruction of global civilization, the only way to keep alive our long tradition of humanistic inquiry is to learn to die. We must practice suspending stress-semantic chains of social excitation through critical thought, contemplation, philosophical debate, and posing impertinent questions. We must suspend our attachment to the continual press of the present by keeping alive the past, cultivating the info-garden of the archive, reading, interpreting, sorting, nurturing, and, most important, reworking our stock of remembrance. We must keep renovating and innovating perceptual, affective, and conceptual fields through recombination, remixing, translation, transformation, and play. We must inculcate ruminative frequencies in the human animal by teaching slowness, attention to detail, argumentative rigor, careful reading, and meditative reflection. We must keep up our communion with the dead, for they are us, as we are the dead of future generations.

Wars begin and end. Empires rise and fall. Buildings collapse, books burn, servers break down, cities sink into the sea. Humanity can survive the demise of fossil-fuel civilization and it can survive whatever despotism or barbarism will arise in its ruins. We may even be able to survive in a greenhouse world. Perhaps our descendents will build new cities on the shores of the Arctic Sea, when the rest of the Earth is scorching deserts and steaming jungles. If being human is to mean anything at all in the Anthropocene, if we are going to refuse to let ourselves sink into the futility of life without memory, then we must not lose our few thousand years of hard-won knowledge, accumulated at great cost and against great odds. We must not abandon the memory of the dead.

As biological and cultural diversity is threatened across the world by capitalist monoculture and mass extinction, we must build arks: not just biological arks, to carry forward endangered genetic data, but also cultural arks, to carry forward endangered wisdom. The library of human cultural technologies that is our archive, the concrete record of human thought in all languages that comprises the entirety of our existence as historical beings, is not only the seed stock of our future intellectual growth, but its soil, its source, its womb. The fate of the humanities, as we confront the end of modern civilization, is the fate of humanity itself.

COMING HOME

When the last strip of light is dimming
When the spotlight starts to fade
If there's no tomorrow
You better live

—Sleater-Kinney, "Fade"

The universe as we understand it began in a burst of light, and all we know came to being through the titanic explosions of stars. Slowly galaxies took shape as spinning accretions of rock began to circle individual suns. Our own planet formed out of gas and dust about four and a half billion years ago, and around a billion years after that, deep within Earth's hot, turbulent oceans, life began.[119] Microbial organisms flourished, diversified, and evolved, but within a billion years they faced a climatic transformation that almost sterilized the planet: proliferating photosynthetic cyanobacteria filled the atmosphere with oxygen, which was toxic to the planet's dominant anaerobic life forms and nearly wiped them out. The oxygen also reacted with and reduced atmospheric methane, sending the Earth into a three-hundred-million-year-long ice age. The Great Oxygenation Event dramatically transformed the Earth's biosphere. It also made possible the evolution of more complex plant and animal life.

Over the past two billion years, complex single-celled and then multicellular life has developed in dynamic cycles of population booms and mass extinctions. Individual genetic lines, entire species, and even whole ecosystems emerge, evolve, and disappear as environmental and climatic conditions change. A few billion years in the future, as our sun matures and grows brighter, it will heat the Earth to a super-greenhouse state, annihilating all terrestrial life. Eventually our sun will balloon into a red giant, then collapse. On an even grander scale, the universe will probably either expand entropically into a dim, frozen emptiness of slowly dissipating black holes, or collapse in on itself and explode again, in an endless cycle of crunches and bangs.

This astonishing cosmos is our home. There is no other. There is no Heaven, no Hell, no Judgment, no Elysium. We humans are precocious multicellular energy machines building hives on a rock in space, machines made up of and connected to countless other machines, each of us a microcosm. Trillions and trillions of microorganisms live on our skin and in our stomachs, mouths, intestines, and respiratory tracts while we spin through our lives in innumerable intersecting orbits, shaped and pulled by forces beyond our reckoning. We are machines of machines in machines, all and each seeking homeostatic perpetuation, and our lives and deaths pass through this great cycle like mosquitos rising and falling in a puddle drying in the summer sun. Life, whether for a mosquito, a person, or a civilization, is a constant process of becoming, a continual emergence into patterns of attraction and aversion, desire and suffering, pleasure and pain. Life is a flow. The forms it takes are transient. Death is nothing more than the act of passing from one pattern into another.

Planets do not decide to spin. Stars do not will themselves ablaze. The universe into which we have been born breathes and burns by intricate, mathematical logic, yet while the Earth's formation had causes, it did not have reason or meaning. As Heraclitus wrote: "The universe, which is the same for all, has not been made by any god or man, but it always has been, is, and will be—an ever-living fire, kindling itself by regular measures and going out by regular measures."[120] No hand guided the accretion of gas and dust that formed our sphere, and the dust itself did not intend to become a body. Just so with the microbes born in the Earth's deep seas, the oxygen-spewing cyanobacteria, the first plants and fish and reptiles, the monstrous dinosaurs, the earliest mammals, and the recent efflorescence of primate life. No intention gave rise to *Homo sapiens* and no outside form grants it value. The causality behind our human bloom is the same causality behind rainfall, quasars, and the roll of dice.

Yet as humans evolved complex social networks, language, consciousness, and then culture, we came to organize our lives through systems of belief that saw not merely agency in the world, but will. We saw gods in the weather, in the trees, and in ourselves. When Homer's Greeks stalked the battlefield, Ares drove them in frenzies to kill and Athena stayed their hands. For those ancients, the will of men was subject to the will of gods, and all were ruled by fate. Causality was comprehended by seeing the universe as a web of personified forces. It was only later, after the rise of literacy, that Greek sages and poets began to articulate a difference we take as fundamental today, the distinction between human will and natural force. The independent persistence of written language—*logos*—became the structuring metaphor for the independent persistence of the human mind. We began to believe in the freedom of thought.

As human civilizations rose and fell, devastated by plagues and famine, torn apart by changing climatic conditions, plundered by ravaging hordes, or collapsing under their own inertia, philosophers, priests, and poets kept trying to understand not only *how* the world worked, but what it *meant*. What does it mean to be human? What does it mean to live? What is good? What is truth? These questions plague us. We torture ourselves with them, making ourselves sick, driving ourselves insane. And even as we try to make sense of our lives, we are harried on all sides and at all hours by stimuli, possessed by yearnings and passions we wouldn't choose, don't understand, and can barely control. As we practice interrupting circuits of reaction, however, striving to recognize the forces that work on and through us, we come more and more to see how these forces are the very wiring that connects us to each other and to the universe. As the god Krishna told Arjuna on the eve of battle, in the *Bhagavad-Gita*:

Know that through lucid knowledge
one sees in all creatures
a single, unchanging existence,
undivided within its divisions.[121]

We are finite and limited machines, but we are not merely machines: we are vibrating bodies of energy, condensations of stellar dust and fire, at once matter and life, extension and thought, moment and frequency. The iron in our blood, the oxygen we breathe, and the carbon of which we are composed were all created in the dying hearts of stars. We are creatures of light, and can find in our history the lineaments of a photohumanism going back to ancient days, a form of thought more powerful than any electronic web, more profound than any merely social media. As was written in the Book of Proverbs, "The human spirit is the lamp of God, searching all the innermost parts."[122]

If you look up at the sky on a clear night, far from the city's glow, you will see streaming out of the darkness light that has taken millions, billions of years to reach you. The farthest observed galaxy, photographed by the Hubble Space Telescope, is 13.2 billion light years away.[123] Its photons have been hurtling toward Earth almost since the Big Bang. Its light is so dim that it is invisible to the naked eye. Most of the universe, in fact, cannot be seen unaided: of the 300 sextillion stars astronomers estimate that the known universe contains, you can see, on a dark, clear night, probably no more than a few million.

Our knowledge of this vast universe remains ridiculously limited. At the same time, that same knowledge is utterly awesome. For a growth of carbon scum on a spinning rock in the backwater of an unremarkable galaxy light years from anywhere to develop the technology to send radio telescopes into space to measure the age of the universe is a prodigious achievement. Our primate curiosity and intellectual hubris have inspired breathtaking audacities. Just a few thousand years ago, we were learning to make marks on clay. In the blink of an eye, we've brushed our fingers against eternity.

It may be that we have crossed the summit of our knowledge and power, and the brief explosion of human life in the Holocene will turn out to have been as transient as an algae bloom. It may be, on the other hand, that we'll find a way to survive in the Anthropocene, perhaps even find ways to maintain human civilization in some recognizable form. Whether we survive or not, however, has already been laid out in the explosion of quantum energy that, more than thirteen billion years ago, began the chain of events and reactions that have led to this moment: me writing this page, you reading it.

The universe, to adapt a proposition of Wittgenstein, is everything that is the case.[124] It is total and complete. There is no outside, it lacks nothing, and it cannot be other than what it is. A slight tweak in any variable and everything would be completely different. To have this moment, as you breathe and read, everything that happened for the last thirteen billion years had to happen in exactly the way it did. Not a single atom can have been out of place, not a single muon faulty. Nothing went wrong. No mistakes were made. There was no sin, no error, no fall. There was only necessity.

And so it will be tomorrow, and the day after, and the day after that. If, like a god, we could see every photon's arc and each neutrino's wobble, we would see past and future laid out in a single mathematical design: infinite, determined, perfect.

We will never achieve such knowledge. We only ever see the pattern dimly and in flashes. Yet we can practice and cultivate understanding the intimate, necessary connection of all things to each other. Light comes to us from millions of miles away, through the emptiness of space, and we can see it. Its heat warms our skin. Pleasure arises in feeling ourselves attuned and connected to such sublime power. The only practical question remaining is whether we, existing as we are, will be that light.

SELECTED BIBLIOGRAPHY

Many more books, websites, lectures, and reports were consulted in writing this book than are directly referenced. This brief bibliography, focusing on climate change and some of its philosophical and ethical impacts, names some of these works along with selected works cited. Please see the endnotes for references not listed here.

Archer, David. *The Long Thaw: How Humans Are Changing the Next 100,000 Years of Earth's Climate*. Princeton: Princeton University Press, 2009.

Broome, John. *Climate Matters: Ethics in a Warming World*. New York: W.W. Norton, 2012.

Dyer, Gwynne. *Climate Wars*. Toronto: Random House, 2008.

Fagan, Brian. *The Long Summer: How Climate Changed Civilization*. Cambridge, MA: Basic Books, 2004.

Grubb, Michael, with Jean-Charles Hourcade and Karsten Neuhoff. *Planetary Economics: Energy, Climate Change and the Three Domains of Sustainable Development*. New York: Routledge, 2014.

Hansen, James. *Storms of My Grandchildren: The Truth About the Coming Climate Catastrophe and Our Last Chance to Save Humanity*. New York: Bloomsbury, 2009.

Hertsgaard, Mark. *Hot: Living Through the Next Fifty Years on Earth*. New York: Houghton Mifflin, 2011.

Hulme, Mike. *Why We Disagree About Climate Change: Un-*

derstanding Controversy, Inaction, and Opportunity*. New York: Cambridge, 2009.

Jamieson, Dale. *Reason in a Dark Time: Why the Struggle Against Climate Change Failed—And What It Means for Our Future*. New York: Oxford University Press, 2014.

Klein, Naomi. *This Changes Everything: Capitalism vs. the Climate*. New York: Simon & Schuster, 2014.

Klein, Richard G. *The Human Career: Human Biological and Cultural Origins*, 3rd edition. Chicago: University of Chicago Press, 2009

Kolbert, Elizabeth. *The Sixth Extinction: An Unnatural History*. New York: Henry Holt & Co., 2014.

McKibben, Bill. *Eaarth: Making a Life on a Tough New Planet*. New York: St. Martins, 2011.

Mitchell, Timothy. *Carbon Democracy: Political Power in the Age of Oil*. New York: Verso, 2011.

Monbiot, George. *Heat: How to Stop the Planet Burning*. London: Allen Lane (Penguin), 2006.

Morton, Timothy. *The Ecological Thought*. Cambridge, MA: Harvard University Press, 2010.

Mulgan, Tim. *Ethics for a Broken World: Imagining Philosophy After Catastrophe*. Cambridge: Acumen Publishing, 2011.

Nixon, Rob. *Slow Violence and the Environmentalism of the Poor*. Cambridge, MA: Harvard University Press, 2013.

Norgaard, Kari Marie. *Living in Denial: Climate Change, Emotions, and Everyday Life*. Cambridge, MA: MIT, 2011.

Pielke, Jr., Roger. *The Climate Fix: What Scientists and Politicians Won't Tell You About Global Warming*. New York: Basic Books, 2010.

Scheffler, Samuel. *Death and the Afterlife*. New York: Oxford University Press, 2013.

Smil, Vaclav. *Global Catastrophes and Trends: The Next 50 Years*. Cambridge, MA: MIT, 2008.

Spinoza, Baruch. *The Ethics*. Trans. Samuel Shirley. Indianapolis: Hackett, 1992.

Wilson, E.O. *The Future of Life*. New York: Vintage, 2003.

Yergin, Daniel. *The Quest: Energy, Security, and the Remaking of the Modern World*. New York: Penguin, 2011.

ENDNOTES

1. Bryan Bender, "Chief of US Pacific forces calls climate biggest worry." *Boston Globe*, March 9, 2013. http://www.bostonglobe.com/news/nation/2013/03/09/admiral-samuel-locklear-commander-pacific-forces-warns-that-climate-change-top-threat/BHdPVCLrWEMxRe9IXJZcHL/story.html.

2. "Remarks by Tom Donilon, National Security Advisor to the President At the Launch of Columbia University's Center on Global Energy Policy," April 23, 2014. Press release issued by The White House, Office of the Press Secretary. http://www.whitehouse.gov/the-press-office/2013/04/24/remarks-tom-donilon-national-security-advisor-president-launch-columbia-.

3. James R. Clapper, "Statement for the Record: Worldwide Threat Assessment of the US Intelligence Community." Senate Select Committee on Intelligence, March 12, 2013, 9. http://www.dni.gov/files/documents/Intelligence%20Reports/2013%20ATA%20SFR%20for%20SSCI%2012%20Mar%202013.pdf.

4. The White House, *National Security Strategy 2010*. http://www.whitehouse.gov/sites/default/files/rss_viewer/national_security_strategy.pdf. US Department of Defense, *2014 Quadrennial Defense Review*; US Department of Homeland Security, *2014 Quadrennial Homeland Security Review*.

5. The US Department of Defense. *2014 Climate Change Adaptation Roadmap*. http://www.acq.osd.mil/ie/download/CCARprint.pdf.

6. World Bank, *Turn Down the Heat: Climate Extremes, Regional Impacts, and the Case for Resilience*, Washington, D.C.: International Bank for Reconstruction and Development/The World Bank, 2013. http://www-wds.worldbank.org/external/default/WDSContentServer/WDSP/IB/2013/06/14/000445729_20130614145941/Rendered/PDF/784240WP0Full00D0CONF0to0June19090L.pdf. World Bank, *Turn Down the Heat: Confronting the New Climate Normal*, Washington, D.C.: International Bank for Reconstruction and Development/The World Bank, 2014. http://www-wds.worldbank.org/external/default/

WDSContentServer/WDSP/IB/2014/11/20/000406484_20141120 090713/Rendered/PDF/927040v20WP00O0ull0Report000English. pdf.

7. Sam Carana, "Heatwave to Hit Greenland," *Arctic News*, August 15, 2014. http://arctic-news.blogspot.com/2014/08/heatwave-to-hit-greenland.html.

8. Ian Joughin, Benjamin E. Smith, and Brooke Medley, "Marine Ice Sheet Collapse Potentially Under Way for the Thwaites Glacier Basin, West Antarctica," *Science* 344:6185 (May 16, 2014), 735–738; and Thomas Sumner, "No Stopping the Collapse of West Antarctic Ice Sheet," *Science* 344:6185 (May 16, 2014), 683. James Hansen writes: "Loss of the entire West Antarctic ice sheet would raise sea level 6 to 7 meters (20 to 25 feet) and eventually open a path to the ocean for part of the much larger East Antarctic ice sheet. Once the ice sheets' collapse begins, global coastal devastations and their economic reverberations may make it impractical for humanity to take actions to rapidly reverse climate forcings." James Hansen, *Storms of My Grandchildren: The Truth About the Coming Climate Catastrophe and Our Last Chance to Save Humanity* (New York: Bloomsbury, 2009), 83.

9. John Kessler, "Seafloor methane: Atlantic bubble bath," *Nature Geoscience* 7 (2014), 625–626.

10. Steve Connor, "Vast methane 'plumes' seen in Arctic ocean as sea ice retreats," *The Independent*, December 13, 2011. http://www.independent.co.uk/news/science/vast-methane-plumes-seen-in-arctic-ocean-as-sea-ice-retreats-6276278.html; Katia Moskvitch, "Mysterious Siberian crater attributed to methane," *Nature*, July 31, 2014. http://www.nature.com/news/mysterious-siberian-crater-attributed-to-methane-1.15649; Natalia Shakhova, Igor Semiletov, Anatoly Salyuk, Vladimir Yusupov, Denis Kosmach, and Örjan Gustafsson, "Extensive Methane Venting to the Atmosphere from Sediments of the East Siberian Arctic Shelf," *Science* 327:5970 (March 5, 2010), 1246–1250; Natalia Shakhova, V.A. Alekseev, and I.P. Semiletov, "Predicted Methane Emission on the East Siberian Shelf," *Doklady Earth Sciences* 430:2 (2010), 190–193; "SWERUS-C3: First observations of methane release from Arctic Ocean hydrates," Stockholm University,

http://www.su.se/english/research/leading-research-areas/science/swerus-c3-first-observations-of-methane-release-from-arctic-ocean-hydrates-1.198540; Gail Whiteman, Chris Hope, and Peter Wadhams, "Climate science: Vast costs of Arctic change," *Nature* 499 (July 25, 2013), 401–403; and Sergey A. Zimov, Edward A.G. Schuur, F. Stuart Chapin III. "Permafrost and the Global Carbon Budget," *Science* 312:5780 (June 16, 2006), 1612–1613. Also see Sam Carana at the Arctic News blog, http://arctic-news.blogspot.com/.

11. David Archer, *The Long Thaw: How Humans Are Changing the Next 100,000 Years of Earth's Climate* (Princeton: Princeton University Press, 2009), 132.

12. Intergovernmental Panel on Climate Change, "Climate Change 2014 Synthesis Report," November 5, 2014. http://www.ipcc.ch/news_and_events/docs/ar5/ar5_syr_headlines_en.pdf. This document is the Headline Summary of the IPCC, "Summary for Policymakers," *Climate Change 2014: Mitigation of Climate Change: Working Group III Contribution to the Fifth Assessment Report of the Intergovernmental Panel on Climate Change* (New York: Cambridge University Press, 2014). http://www.ipcc.ch/report/ar5/wg3/.

13. "There is growing evidence, that even with very ambitious mitigation action, warming close to 1.5 degrees Celsius above pre-industrial levels by mid-century is already locked in to the Earth's atmospheric system and climate change impacts such as extreme heat events may now be unavoidable." World Bank, *Turn Down the Heat: Confronting the New Climate Normal*, xvii.

14. Camilo Mora, et al. "The projected timing of climate departure from recent variability," *Nature* 502 (October 10, 2013), 183–187. 186.

15. James Hansen, "Game Over for the Climate," *New York Times*, May 9, 2012. http://www.nytimes.com/2012/05/10/opinion/game-over-for-the-climate.html.

16. "The term Anthropocene suggests: (i) that the Earth is now moving out of its current geological epoch, called the Holocene and (ii) that human activity is largely responsible for this exit from the Holocene, that is, that humankind has become a global geological force in its own right." Will Steffen, Jacques Grinevald, Paul Crutzen, and John Mc-

Neill, "The Anthropocene: conceptual and historical perspectives," *Philosophical Transactions of the Royal Society* 369: 1938 (March 2011), 842–867.

17. Paul J. Crutzen and Eugene F. Stoermer, "The Anthropocene," *Global Change Newsletter: International Geosphere–Biosphere Programme Newsletter* 41 (May 2000), 17–18.

18. William Blake, *Milton*, f. 20, ll. 15–25, *The Poetical Works of William Blake*, ed. John Sampson (London: Oxford University Press, 1908).

19. Colin N. Waters, Jan A. Zalasiewicz, Mark Williams, Michael A. Ellis and Andrea M. Snelling, "A stratigraphical basis for the Anthropocene?" Geological Society, London, Special Publications 395 (2014), 1–21. Jan Zalasiewicz, et al., "Stratigraphy of the Anthropocene," *Philosophical Transactions of the Royal Society* 369:1938 (March 2011), 1036–1055.

20. Leading researchers on the subject now argue that this date would serve as the best "golden spike" by which to mark the beginning of the Anthropocene. Jan Zalasiewicz, et al., "When did the Anthropocene begin? A mid-twentieth century boundary level is stratigraphically optimal," *Quaternary International* (forthcoming, online January 12, 2015). doi:10.1016/j.quaint.2014.11.045. http://www.sciencedirect.com/science/article/pii/S1040618214009136.

21. As Ian Baucom has recently written, "Although I have for some time accepted the force of Fredric Jameson's dictum that 'we cannot, not periodize,' until very recently it would not have occurred to me that postcolonial study, critical theory, or the humanities disciplines in general needed to periodize in relation not only to capital but to carbon, not only in modernities and post-modernities but in parts-per-million, not only in dates but in degrees Celsius." Ian Baucom, "History 4°: Postcolonial Method and Anthropocene Time," *Cambridge Journal of Postcolonial Literary Inquiry* 1:1 (March 2014), 123–142. 125.

22. Dale Jamieson. *Reason in a Dark Time: Why the Struggle Against Climate Change Failed—And What It Means for Our Future* (New York: Oxford University Press, 2014), 185.

23. Dipesh Chakrabarty, "Postcolonial Studies and the Challenge of Climate Change," *New Literary History* 43:1 (Winter 2012), 1–18. 9.

24. Michel de Montaigne, *Essays and Selected Writings*, trans. Donald M. Frame (New York: St. Martin's Press, 1963), 9.

25. Shout out to Smokey and Stretch, C Battery, 1/94FA medics.

26. Simone Weil, "The Iliad, or the Poem of Force," *War and the Iliad*, trans. Mary McCarthy (New York: The New York Review of Books, 2005), 22.

27. Yamamoto Tsunetomo, *Hagakure: The Book of the Samurai*, trans. William Scott Wilson (New York: Kodansha International, Ltd., 1979), 164. It was Jim Jarmusch's spooky, beautiful film *Ghost Dog: Way of the Samurai* (1999) that brought Tsunetomo to my attention.

28. Peter Sloterdijk, "Rules for the Human Zoo: a response to the *Letter on Humanism*," trans. Mary Varney Rorty, *Environment and Planning D: Society and Space* 27 (2009), 12–28. 24.

29. MS Harley 2253, f.59v, British Library, London. http://www.bl.uk/catalogues/illuminatedmanuscripts/ILLUMIN.ASP?Size=mid&IllID=19740. Thanks to D. Vance Smith for introducing me to this haunting lyric.

30. Richard G. Klein, *The Human Career: Human Biological and Cultural Origins*, 3rd edition (Chicago: University of Chicago Press, 2009), 615.

31. This happens because of variations in the orbital parameters of the Earth—the way the planet slowly wobbles as it spins around the sun.

32. Klein, *The Human Career*, 645.

33. Ibid., 656–659.

34. This may be happening again today as glacial meltwater from Greenland and the Arctic floods the North Atlantic. See Stefan Rahmstorf, et al., "Exceptional twentieth-century slowdown in Atlantic Ocean overturning circulation," *Nature Climate Change* (March 23, 2015). http://www.nature.com/nclimate/journal/vaop/ncurrent/pdf/nclimate2554.pdf.

35. Here and throughout this chapter I rely heavily on Brian Fagan, *The Long Summer: How Climate Changed Civilization* (Cambridge, MA: Basic Books, 2004).

36. Fagan, 141–145. See also H.M. Cullen, P. B. deMenocal, S. Hem-

ming, G. Hemming, F. H. Brown, T. Guilderson and F. Sirocko, "Climate change and the collapse of the Akkadian empire: Evidence from the deep sea," *Geology* 28 (2000), 379–382; and H. Weiss, M.A. Courty, W. Wetterstrom, F. Guichard, L. Senior, R. Meadow and A. Curnow, "The Genesis and Collapse of Third Millennium North Mesopotamian Civilization," *Science*, New Series 261: 5124 (August 20, 1993), 995–1004. Traditional political explanations for the collapse of the Akkadian empire describe an invasion by the "Guti" or "Gutium" people. As Georges Roux writes: "About the Guti who overthrew the Akkadian empire and ruled over Mesopotamia for almost a hundred years, we know next to nothing. The Sumerian King List gives 'the hordes of Gutium' twenty-one kings, but only five of them have left us inscriptions and this, coupled with silence from other sources, points to a period of political unrest and cultural regression." Georges Roux, *Ancient Iraq* (London: George Allen & Unwin Ltd., 1964), 155. The period remains hazy: "Different dates are given for the end of the dynasty of Akkad. . . . It is unclear how long this 'intermediate' period lasted, and because of the (almost complete) absence of written records it is difficult to reconstruct events between the Akkad dynasty and the Third Dynasty of Ur." Nicole Brisch, "History and chronology," *The Sumerian World*, ed. Harriet Crawford (New York: Routledge, 2013), 122.

37. See US Environmental Protection Agency, "Atmospheric Concentrations of Greenhouse Gases," figure 2. http://www.epa.gov/climatechange/science/indicators/ghg/ghg-concentrations.html.

38. Hansen, *Storms*, 50.

39. Archer, 130.

40. Ibid., 95.

41. V. Sissakian, N. Al-Ansari, and S. Knutsson, "Sand and dust storm events in Iraq," *Natural Science* 5:10 (2013), 1084–1094.

42. Svante Arrhenius, "On the Influence of Carbonic Acid in the Air upon the Temperature of the Ground," *The London, Edinburgh, and Dublin Philosophical Magazine and Journal of Science*, Fifth Series 41:251 (April 1896). Extracted from a paper presented to the Swedish Royal Academy of Sciences, December 11, 1895.

43. Philip Shabecoff, "Global Warming Has Begun, Expert Tells

Senate," *The New York Times*, June 24, 1988. http://www.nytimes.com/1988/06/24/us/global-warming-has-begun-expert-tells-senate.html.

44. US Environmental Protection Agency, "Climate Change Indicators in the United States: Global Greenhouse Gas Emissions" (updated May 2014). http://www.epa.gov/climatechange/science/indicators/ghg/global-ghg-emissions.html.

45. Intergovernmental Panel on Climate Change, Working Group III, "Chapter 7: Energy Systems," *Climate Change 2014: Mitigation of Climate Change: Working Group III Contribution to the Fifth Assessment Report of the Intergovernmental Panel on Climate Change* (New York: Cambridge University Press, 2014), 527. http://www.ipcc.ch/pdf/assessment-report/ar5/wg3/ipcc_wg3_ar5_chapter7.pdf.

46. "[E]stimates from models considered for the IPCC AR4 for stabilization at 550 ppm CO_2eq range from small GDP gains to 4% GDP losses in 2050. For a comparable category (500–550 ppm CO_2eq), the Stern Review found a range of -2 to 5% GDP losses." Ottmar Edenhofer, et al., *RECIPE (Report on Energy and Climate Policy in Europe): The Economics of Decarbonization* (2009). http://www.pik-potsdam.de/members/edenh/publications-1/recipe_report.pdf.

47. Roger Pielke Jr. calls this the "iron law" of climate policy: "When policies focused on economic growth confront policies focused on emissions reductions, it is economic growth that will win out every time." Roger Pielke Jr., *The Climate Fix: What Scientists and Politicians Won't Tell You About Global Warming* (Basic Books: New York 2010), 46.

48. Intergovernmental Panel on Climate Change, "Summary for Policymakers," *Climate Change 2014: Mitigation of Climate Change*.

49. See, for example, Nafeez Ahmed, "IPCC reports 'diluted' under 'political pressure' to protect fossil fuel interests," *The Guardian*, May 15, 2014. http://www.theguardian.com/environment/Earth-insight/2014/may/15/ipcc-un-climate-reports-diluted-protect-fossil-fuel-interests; Dimitri Zenghelis, "Richard Tol's flawed claims about the Stern Review," *Business Spectator*, April 9, 2014. http://www.businessspectator.com.au/article/2014/4/9/policy-politics/richard-tols-flawed-claims-about-stern-review; Biofuelwatch, et al., "IPCC

support for 'false solutions' denounced by climate justice activists," joint press release, April 13, 2014. http://www.biofuelwatch.org. uk/2014/ipccwg3-joint-pr/.

50. Vaclav Smil, *Global Catastrophes and Trends: The Next 50 Years* (Cambridge, MA: MIT, 2008), 82.

51. "Substituting one form of energy for another takes a long time." Vaclav Smil, "A Skeptic Looks at Alternative Energy," *IEEE Spectrum* June 28, 2012. http://spectrum.ieee.org/energy/renewables/a-skeptic-looks-at-alternative-energy/0. See also Arnulf Grübler, Nebošja Nakićenović, and David G. Victor, "Dynamics of energy technologies and global change," *Energy Policy* 27 (1999), 247–280.

52. Gwynne Dyer, *Climate Wars* (Toronto: Random House, 2008), 134.

53. Germany has recently, to much fanfare, produced as much as 27 percent of its electricity from renewable sources, including a mix of photovoltaic, wind, hydro-electric, and bioenergy. But this may come at a cost, which gives worry to economists and energy analysts as German industrial energy prices keep climbing. According to Daniel Yergin, "Germany's current path of increasingly high-cost energy will make the country less competitive in the world economy, penalize Germany in terms of jobs and industrial investment, and impose a significant cost on the overall economy and household income." Matthew Karnitschnig, "Germany's Expensive Gamble on Renewable Energy," *Wall Street Journal*, August 26, 2014. http://www.wsj.com/articles/germanys-expensive-gamble-on-renewable-energy-1409106602.

54. Dyer, 136. For Dyer's source data, see George Monbiot, *Heat: How to Stop the Planet Burning* (London: Allen Lane (Penguin), 2006), 100–123.

55. Hansen, *Storms*, 193.

56. James Hansen, Ken Caldeira, Kerry Emanuel, and Tom Wigley, "Top climate change scientists' letter to policy influencers," CNN, November 3, 2013. http://www.cnn.com/2013/11/03/world/nuclear-energy-climate-change-scientists-letter/.

57. Pielke, 116.

58. Michael Grubb, Jean-Charles Hourcade, and Karsten Neuhoff, *Planetary Economics: Energy, Climate Change and the Three Domains of Sustainable Development* (New York: Routledge, 2014), 28.

59. Ibid., 29.

60. For one account of the history of cap-and-trade, or allowance trading, as it developed in response to sulfur dioxide (SO_2) pollution ("acid rain"), see Daniel Yergin, *The Quest: Energy, Security, and the Remaking of the Modern World* (New York: Penguin, 2011), 475–483.

61. Thomas C. Schelling, "What Makes Greenhouse Sense?" *Strategies of Commitment and Other Essays* (Cambridge, MA: Harvard University Press, 2006), 27–44. 36. Thanks to Deak Nabers for bringing Schelling's work to my attention.

62. Ibid., 36–38.

63. These numbers are from the IEA's 2013 report. International Energy Agency, *Technology Roadmap: Carbon Capture and Storage*, 2013. 10–11. http://www.iea.org/publications/freepublications/publication/TechnologyRoadmapCarbonCaptureandStorage.pdf

64. C. Le Quéré, et al. "Global carbon budget 2013." *Earth System Science Data* 6 (2014), 235–263. 253. http://www.Earth-syst-sci-data.net/6/235/2014/essd-6-235-2014.pdf. Writing in the *Wall Street Journal*, Keith Johnson laid out some of the most glaring problems with the IEA's 2009 Roadmap: "The report starts off talking of $2.5 trillion to $3 trillion in 'additional investment' through 2050. But the report throws trillion-dollar figures around with such abandon, it's hard to measure the true cost. A few pages later, for instance, the report estimates the 'additional cost' of all the carbon capture projects in the world at $5.8 trillion. Then there is the sheer physical difficulty of installing 3,400 carbon-capture projects around the world by 2050: That's an average of 85 projects per year, every year, till the middle of the century, or one every four days. Starting pitchers can't even keep up that pace to throw a few innings; imagine trying to make, move, install, test, and commission large-scale carbon-capture projects at that pace. And then find a place to put all those millions of tons of carbon dioxide. Because carbon capture without storage is meaningless. The carbon emissions that are caught have to get stuffed underground via

pipelines. The IEA figures 360,000 kilometers of pipeline should do the trick. That's nine trips around the earth." Keith Johnson, "Catch Me If You Can: Does the IEA's Carbon Capture Plan Make Any Sense?" *The Wall Street Journal*, October 14, 2009. http://blogs.wsj.com/environmentalcapital/2009/10/14/catch-me-if-you-can-does-the-ieas-carbon-capture-plan-make-any-sense/.

65. International Energy Agency, *Technology Roadmap: Carbon Capture and Storage*, 21.

66. Pielke, 135.

67. As Schelling points out, our use of the term "geoengineering" is peculiar: "'Geoengineering' implies something unnatural. I would suppose, for example, that if the Earth's atmosphere had always had a large amount of sulfur aerosols in the upper atmosphere and the aerosols increased and diminished from time to time and the carbon dioxide increased and diminished from time to time, and we began to have a greenhouse problem, it would be referred to as an imbalance in the ratio of the infrared-absorbing substances to the light-reflecting substances; reducing CO_2 and increasing the sulfur would both appear unnatural. If we put carbon black on the Arctic ice to make it disappear, that would be considered geoengineering; if we just let it disappear because of global warming, that is not geoengineering. If we learn to make it snow more in the Sierras and the Rockies to enhance the water supply of California and Colorado and improve the ski slopes in the winter, that is not geoengineering; if we learn to make it snow in Antarctica, in order to store water there to reduce the sea level, that is geoengineering." Thomas Schelling, "The Economic Diplomacy of Geoengineering," *Strategies of Commitment and Other Essays* (Cambridge, MA: Harvard University Press, 2006), 45–50.

68. Most notably by Mike Hulme. Mike Hulme, *Why We Disagree About Climate Change: Understanding Controversy, Inaction, and Opportunity* (New York: Cambridge, 2009), 334–337.

69. For a detailed account of the following process, see Thomas Seeley, *Honeybee Democracy* (Princeton: Princeton University Press, 2010).

70. Timothy Mitchell, *Carbon Democracy: Political Power in the Age of Oil* (New York: Verso, 2011), 12–42.

71. Mitchell, 27.

72. Mitchell, 39.

73. See, among others, Arun Gupta, "How the People's Climate March Became a Corporate PR Campaign," *Counterpunch*, September 19–21, 2014. http://www.counterpunch.org/2014/09/19/how-the-peoples-climate-march-became-a-corporate-pr-campaign/

74. NYPD Detective Rick Lee, the "Hipster Cop" of Occupy Wall Street fame.

75. Tang Jie's powerpoint presentation is available here: https://ieta.memberclicks.net/assets/UNFCCC/New_York_2014/Presentations/2014922ny.pdf.

76. Daniel Bodansky, Seth Hoedl, Gilbert Metcalf, and Robert Stavins, "Facilitating Linkage of Heterogeneous Regional, National, and Sub-National Climate Policies Through a Future International Agreement." http://belfercenter.ksg.harvard.edu/files/ieta-hpca-essept2014.pdf.

77. As the executive summary reports: "Although the negotiations are still at a relatively early stage, it appears likely that the 2015 agreement will reflect a hybrid climate-policy architecture—one that combines top-down elements, such as for measurement (or monitoring), reporting, and verification (MRV), with bottom-up elements consisting primarily of 'nationally determined contributions' (NDCs). In their NDCs, countries would specify their own targets, actions, policies—or some combination of these—to reduce greenhouse-gas emissions. The character and ambition of NDCs will be based upon domestic political feasibility and other factors, and will be subject to some system of international peer review."

78. http://www.globalcarbonproject.org/carbonbudget/14/hl-full.htm#regionalFF

79. Mark Mills notes: "The information economy is a blue-whale economy with its energy uses mostly out of sight. Based on a mid-range estimate, the world's Information-Communication-Technologies ecosystem uses about 1,500 TWh of electricity annually, equal to all the electric generation of Japan and Germany combined—as much electricity as was used for global illumination in 1985. The ICT eco-

system now approaches 10% of world electricity generation. Or in other energy terms—the zettabyte era already uses about 50% more energy than global aviation." Mark Mills, *The Cloud Begins with Coal: Big Data, Big Networks, Big Infrastructure, and Big Power*, National Mining Association, August 2013, 3. http://www.tech-pundit.com/wp-content/uploads/2013/07/Cloud_Begins_With_Coal.pdf?c761ac. Also see Gary Cook, *How Clean Is Your Cloud?* Greenpeace, April 2012. http://www.greenpeace.org/international/Global/international/publications/climate/2012/iCoal/HowCleanisYourCloud.pdf.

80. Charles E. Cobb, Jr. *This Nonviolent Stuff'll Get You Killed: How Guns Made the Civil Right Movement Possible* (New York: Basic Books, 2014).

81. Cobb, 111.

82. Historian Walter Rucker, discussing Williams's legacy, writes: "While black militancy may indeed date back to the colonial era . . . the emergence of Williams and Malcolm X represented a significant re-awakening of this spirit. What followed in the decade after . . . were waves of militant black revolutionaries, dozens of urban rebellions, and numerous calls for armed self-defense. After centuries of anti-black violence, African-Americans across the country began to defend their communities aggressively—employing overt force when necessary." Walter Rucker, "Crusader in Exile: Robert F. Williams and the International Struggle for Black Freedom in America," *The Black Scholar* 36:2–3 (2006), 19–34. 22.

83. James Baldwin, *The Fire Next Time* (New York: Vintage, 1993 1963), 77.

84. Phillip L. Walker, "A Bioarchaeological Perspective on the History of Violence," *Annual Review of Anthropology* 30 (2001), 586. http://arjournals.annualreviews.org/doi/pdf/10.1146/annurev.anthro.30.1.573.

85. Ibid., 590.

86. Sigmund Freud, "Why War?" *The Standard Edition of the Complete Psychological Works of Sigmund Freud*, Vol 22: (1932-1936): *New Introductory Lectures on Psycho-Analysis and Other Works*, 204. http://www.pep-web.org/document.php?id=SE.022.0195A.

87. Heraclitus, fragment 26. Philip Wheelwright, *Heraclitus* (Princeton: Princeton University Press, 1959), 29.

88. Tom Pyszczynski, "What Are We So Afraid Of? A Terror Management Theory Perspective on the Politics of Fear," *Social Research* 71:4 (Winter 2004), 827–847. 827.

89. Lauren Berlant, *Cruel Optimism* (Durham: Duke University Press, 2011).

90. Francesco Femia and Caitlin Werrell, "Syria: Climate Change, Drought and Social Unrest," The Center for Climate and Security, February 29, 2012. http://climateandsecurity.org/2012/02/29/syria-climate-change-drought-and-social-unrest/.

91. Colin P. Kelley, Shahrzad Mohtadi, Mark A. Cane, Richard Seager, and Yochanan Kushnir, "Climate change in the Fertile Crescent and implications of the recent Syrian drought," *Proceedings of the National Academy of Sciences of the United States of America*, January 30, 2015. Early edition. http://www.pnas.org/content/early/2015/02/23/1421533112.

92. Quoted in Eric Holthaus, "New Study Says Climate Change Helped Spark Syrian Civil War," *Slate.com*, March 3, 2015. http://www.slate.com/blogs/future_tense/2015/03/02/study_climate_change_helped_spark_syrian_civil_war.html.

93. International Committee for the Red Cross, "Gaza: Water in the line of fire," news release, July 15, 2014. http://www.icrc.org/eng/resources/documents/news-release/2014/14-07-israel-palestine-gaza-water.htm. Ahmed Hadi, "Health crisis looms in Gaza after Israel bombs water infrastructure," *Al-Akhbar English*, July 17, 2014. http://english.al-akhbar.com/content/health-crisis-looms-gaza-after-israel-bombs-water-infrastructure

94. Nafeez Ahmed, "IDF's Gaza assault is to control Palestinian gas, avert Israeli energy crisis," *The Guardian*, July 9, 2014. http://www.theguardian.com/environment/Earth-insight/2014/jul/09/israel-war-gaza-palestine-natural-gas-energy-crisis. Julie Lévesque, "Israel Steals Gaza's Offshore Natural Gas: $15 Billion Deal with Jordan," *Global Research*, September 06, 2014. http://www.globalresearch.ca/israel-steals-gazas-offshore-natural-gas-15-billion-deal-with-jordan/5399736.

95. Jeff Wilson, "Ukraine's Wheat, Corn Face Mounting Drought Risk, Martell Says," *Bloomberg.com*, March 5, 2014. http://www.bloomberg.com/news/2014-03-05/ukraine-s-wheat-corn-face-mounting-drought-risk-martell-says.html.

96. Richard Heede, "Tracing anthropogenic carbon dioxide and methane emissions to fossil fuel and cement producers, 1854–2010," *Climatic Change* (2014) 122:229–241.

97. Peter Sloterdijk, *Neither Sun nor Death* (New York: Semiotext(e), 2011), 84–85.

98. Andrew George, trans., *The Epic of Gilgamesh: The Babylonian Epic Poem and Other Texts in Akkadian and Sumerian* (London: Allen Lane, 1999), 71.

99. See Ernest Becker, *The Denial of Death* (New York: Free Press, 1973), as well as Norman O. Brown, Otto Rank, Robert Jay Lifton, and more recently terror management theory, developed by Thomas Pyszczynski, Sheldon Solomon, and Jeff Greenberg. See, for example, Tom Pyszczynski, "What Are We So Afraid Of?"

100. Becker, *The Denial of Death*, ix.

101. James Jones, *The Thin Red Line* (New York: Dial, 2012), 309.

102. Michel de Montaigne, *Essays and Selected Writings*, 9.

103. "Ordinary people seem not to realize that those who really apply themselves in the right way to philosophy are directly and of their own accord preparing themselves for death and dying." Plato, *Phaedo* 64–68, *Plato: The Collected Dialogues*, ed. Edith Hamilton and Huntington Cairns (Princeton: Princeton University Press, 2005), 46–50.

104. Baldwin, *The Fire Next Time*, 92.

105. Dōgen Zenji, "A Universal Recommendation for Zazen," *Zen Sourcebook: Traditional Documents from China, Korea, and Japan*, ed. Stephen Aldiss (Indianapolis: Hackett, 2008), 143.

106. Marcus Aurelius, *The Thoughts of the Emperor M. Aurelius Antoninus*, II.17, trans. George Long (Boston: Willard Small, 1889), 28.

107. G.W.F. Hegel, "The Philosophy of Spirit," in Leo Rauch, *Hegel and the Human Spirit: A Translation of the Jena Lectures on the Philosophy of Spirit (1805–06) with commentary* (Detroit: Wayne State University

Press, 1983), 87.

108. There is some evidence that Neanderthals developed basic culture through "the occasional discovery of artifact assemblages that comprise a blend of Neanderthal/Mousterian and Cro-Magnon/Upper Paleolithic artifact types." Klein, *The Human Career*, 654.

109. Martin L. West, *Studies in the Text and Transmission of the Iliad* (Munich: K.G. Saur, 2001), 140. I rely on West for his account of the *Iliad*'s survival and early transmission, 5–7.

110. Amy Hackney Blackwell, "Robot Scans Ancient Manuscript in 3-D," *Wired*, June 5, 2007. http://archive.wired.com/gadgets/miscellaneous/news/2007/06/iliad_scan.

111. At the Homer Multitext Project: http://www.homermultitext.org/manuscripts-papyri/venetusA.html.

112. Hannah Arendt, *On Revolution* (New York: Penguin, 2006 1963), 212.

113. See Andrew George, Introduction to *The Epic of Gilgamesh* (London: Allen Lane, 1999), xvi–xxx.

114. George, *The Epic of Gilgamesh*.

115. George, Introduction, *Gilgamesh*, l.

116. What follows relies on reporting I did for *Rolling Stone* magazine. Roy Scranton, "An Iraqi Band's (Semi) Happy Ending," *Rolling Stone*, April 9, 2015.

117. For the story of Acrassicauda, I rely on personal interviews, in addition to the following: *Heavy Metal in Baghdad*, DVD, directed by Suroosh Alvi (Brooklyn: Vice Films, 2007); Andy Capper and Gabi Sifre, *Heavy Metal in Baghdad: The Story of Acrassicauda* (New York: Pocket Books, 2009); http://acrassicauda.com/bio.

118. Jaron Lanier, "Digital Maoism: The Hazards of the New Online Collectivism," *Edge*, May 29, 2006. http://edge.org/conversation/digital-maoism-the-hazards-of-the-new-online-collectivism.

119. Fossil evidence found in Australia shows microbial life on Earth at least 3.48 million years ago. Nora Noffke, Daniel Christian, David Wacey, and Robert M. Hazen, "Microbially Induced Sedimentary Structures Recording an Ancient Ecosystem in the ca. 3.48 Billion-

Year-Old Dresser Formation, Pilbara, Western Australia," *Astrobiology* 13:12 (December 2013), 1103–1124.

120. Heraclitus, fragment 29. Wheelwright, 37.

121. Barbara Stoler Miller, trans., *The Bhagavad-Gita: Krishna's Counsel in Time of War* (New York: Bantam, 1986), 138.

122. Proverbs 20:27.

123. http://www.space.com/17755-farthest-universe-view-hubble-space-telescope.html.

124. Ludwig Wittgenstein, *Tractatus Logico-Philosophicus*, trans. C.K. Ogden (New York: Harcourt, Brace, & Co., 1922), 25.

ACKNOWLEDGMENTS

This book is the result of countless conversations and intellectual encounters, and it wouldn't have happened without the help, advice, and support of quite a few people. Most immediately, of course, thanks are due to my editor at City Lights, Greg Ruggiero, who saw the potential in the *New York Times* essay this book grew from and who nurtured some of the most important ideas that developed out of it. Second, my great thanks to my editors at the *New York Times'* philosophy blog, the incomparable Peter Catapano and the bedazzling Simon Critchley. I am deep in their debt not only for the opportunity they gave me but also for their thoughts, their wisdom, and their support. Special thanks as well go to Ian Baucom, Amanda Anderson, Jane Bennett, and my fellow participants at the School of Criticism and Theory at Cornell University in the summer of 2013, who opened my eyes to the philosophical problems the Anthropocene poses and who helped me begin to think about how we might address them.

Much of the introduction and parts of other chapters were first published in the *New York Times*, and are reproduced here with grateful acknowledgment. Portions of this book were published previously in different form in *Theory & Event* and *Rolling Stone* magazine. I would like to thank Deborah Blum, editor of *Best American Science and Nature Writing 2014* (Mariner Books), who chose my essay to appear in that volume, and Tim Folger, series editor of the *Best American Science and Nature Writing*.

I'm very lucky to know a lot of smart, talented people with the patience and generosity to entertain my emerging ideas and read my rough drafts. The attentive care of such friends and colleagues helped me immensely in thinking more deeply about every aspect of this book, while transforming what could have been a rather lonely meditation into a series of invigorating conversations.

I especially want to thank Laura Carver, Meehan Crist, Johann Frick, Dale Jamieson, Naomi Klein, Deak Nabers, Michael Oppenheimer, and Jacob Siegel for conversation and inspiration. Patrick Blanchfield, Maria DiBattista, Jeff Dolven, Raphael Krut-Landau, Hilary Plum, Dorothea von Moltke, Jason Daniel Schwartz, and Susan Stewart all generously took time to read drafts of the manuscript and gave me important feedback. Special thanks go to Andrew Cole, Jedi of the dialectical arts, and to Martin Woessner, my mentor in the study of civilizational collapse ever since I took his "Cities Under Siege" course at the New School in 2007, both of whom read drafts of the book but whose help and influence on my thought go far deeper than any one text might show.

Parts of this book were delivered as presentations to various audiences, and I am grateful to both the participants and organizers of those events. Namely, I want to thank Dominic Boyer and the Center for Energy and Environmental Studies in the Human Sciences for inviting me to the 4th Annual Cultures of Energy Conference at Rice University, which (unusual for academic conferences) felt like coming home to an intellectual community I hadn't dreamed existed; Suzanne Anker at the School of Visual Arts in Manhattan for inviting me to participate in the Naturally Hypernatural: Visions of Nature conference; Stephanie Wakefield and 1882 Woodbine in Ridgewood, Queens, for inviting me to speak there; and Katie Holten and Dillon Cohen for inviting me to their series of salons discussing the Anthropocene.

I would be remiss if I did not thank City Lights' wise and elegant publisher, Elaine Katzenberger, and its painstaking publicist, Stacey Lewis. Thank you both for making this book what it is. Thanks as well to Nate Dorward, for her peerless proofing. I should also thank my agents, Melissa Chinchillo, Sylvie Greenberg, and Donald Lamm at Fletcher & Co., for their assistance with the manuscript and with the work of publishing it. Their guidance and help have been essential.

My thanks also go to Princeton University and the Mrs. Giles Whiting Foundation, for funding my research on this and other topics. Thanks as well to Rosalind Parry, whose hospitality gave me the time and space in the woods of Maine to finish this book's first draft. Also thanks to Milou, who kept the coyotes away.

As this is my first book, I would ask indulgence to take a moment to thank a few people who have served as teachers, motivators, friends, and sometimes guardian angels along the way, without whom I would never have made it this far: Julie Fox, Christopher Hitchens, James Miller, Melissa Monroe, Ben Prochazka, Faith Purvis, Michael Reinbold, Chris Roberts (who introduced me to Spinoza), and Susan Wolfson. Thank you all.

Finally, and most important, my gratitude to my best editor, my most critical reader, and my most tenacious and unrelenting interlocutor, Sara Marcus. Without her tough patience, dogged hope, and tender care, there would have been little reason for me to write through the despair that confronting catastrophic climate change induces. Any light this book might give off is only a reflection of her love.

ABOUT THE AUTHOR

Roy Scranton's journalism, essays, and fiction have been published in *Rolling Stone*, the *New York Times*, *LIT*, *Boston Review*, *Prairie Schooner*, *Los Angeles Review of Books*, *Contemporary Literature*, *Theory & Event*, *The Appendix*, and elsewhere. He is one of the editors of *Fire and Forget: Short Stories from the Long War* (Da Capo, 2013). He grew up in Oregon, earned a Bachelor's degree and a Master's degree at the New School for Social Research, and is completing a PhD in English at Princeton University.